FORCE

OF

RECKONING

JEFFERY L CHENEY

CRAIG J CHENEY

JARED L CHENEY

FORCE OF RECKONING

Published by 7Cs Books, LLC,
P.O. Box 231
Vernonia, OR 97064
www.7CsBooks.com

DEDICATION

To our mother, for a love of reading, and to our father, for a love of learning.

ACKNOWLEDGEMENT

A big thank you to Hans de Ridder for allowing the use of his ship designs in our cover art! Thanks also to our wives for putting up with the time spent writing, rewriting, editing, rewriting some more, and then some more editing, right before the rewriting...

CHAPTER 1

Antoc System
5-8 August 2787

Major Sheli Chowdhury sprinted for the open hatch of the ship. Everyone else should be on board by now. She had seen Mackey and Long make it in. While Long might be a worthless pile of maggots, Mackey wouldn't leave anyone out there to be shot. She had made sure Mitchell and Hayes had gotten clear before she started back. Those two Ensigns didn't have the sense to come in out of the rain, so she had to make sure they were headed for the ship before she left her position to cover their exit.

And then she'd run into those two DaGaman Marines on her way. She'd gotten one of them but the big one, the one Roberts had hit with the big rock back on A3, he'd gotten away.

She was pretty sure she'd clipped him though. He'd gone down momentarily and dropped his rocket pack before retreating. If he'd stuck around, he'd end up dead like all of his buddies. He was unsupported where he was, so retreat was a smart move on his part.

She knew the feeling. Even though he had botched his dawn assault, she too was unsupported, and that had her and her team running, again. She'd wished that she'd gotten him though. He acted like he was the leader. She definitely wanted payback. That was why she was carrying his rocket pack. She was going to return those rockets to him, the first chance she got. Maybe not the pack, but definitely the rockets.

That was for later. For now, she ran for all she was worth as she watched the engines begin to fire on *Vanguard*. It wouldn't do to get left behind.

She cleared out of the brush at the edge of the woods and sprinted across the last few meters to the hatch.

A quick glance showed her Mackey, Long and both ensigns on the floor in front of her.

"All clear, Captain. Lift ship!"

She grabbed for a handhold and searched for something to strap herself down with. They had long ago stripped the ship of any seats or other amenities, so she braced herself as well as she could. This was aided by the fact that there were so many bodies jammed into the small space that no one could move very far.

As soon as she was seated, the engines rumbled more deeply and the ship lurched and began its ascent. Chowdhury was pushed against her captured rocket pack and the brutal acceleration caused it to dig into her ribs, more exposed than usual with the weeks of short rations. The ship's powerful in-system engines took the small craft into orbit quickly, so she didn't have long to curse her decision to grab the weapon.

After lift-off, Captain William Brighton dropped the acceleration down to 1 g and started them on the last leg of their journey.

They were the nineteen, eighteen after A3, loyal crewmembers of the Warner Navy Ship *Pathfinder*. They had spent 38 days getting to this point in their attempt to escape the Antoc System and return to Earth. They had been victims of a mutiny, kicked off their own ship at gunpoint, shot at by missiles, attacked on two separate planets by Marines of another Family, hunted by wild animals, lived on the verge of starvation, and wandered from planet to planet in search of the supplies they needed to get home. But Captain Brighton had refused to let them give up. And now they were almost there.

Captain Brighton's best friend, Edward Teach, had been the leader of the mutineers, and he had disabled the launch they now traveled in by removing the batteries that were used to power the in-system engines and the jump drive, leaving them only enough battery power to reach and land on the nearest planet. He didn't know that Brighton had been part of the expedition that had originally surveyed this system and knew of supply caches on various planets. They had used that knowledge to retrieve the batteries they needed to escape the system and return to Earth. But they still might all die before they got there. The reason *Pathfinder* had been worth stealing was that it had revolutionary new engines that would make the jump gates obsolete. The launch they were in, *Vanguard*, had those same engines; it had been an earlier prototype and testbed for the engines. Apparently, those hostile Marines knew that and would stop at nothing to get their hands on *Vanguard*, now that *Pathfinder* had slipped through their fingers.

The captain soon turned over the piloting duties to Lieutenant Fyonna Johnson and began mapping the gravitation readings in as much detail as possible. He would need to have exact readings to be able to localize the jump point that would enable them to escape from the system.

From her position, Chowdhury could hear the conversation in the cockpit. She shook her head as she remembered Fujinami telling her that the ensigns had taken to calling the cockpit 'Brighton's office.' That thought made her search for the diminutive exobiologist. He was not in the ship. Had she left him on the planet in her haste? Remembering that the Captain had sent Mackey to help him to the ship after the battle, she turned to the burly NCO who was sitting next to her.

"Where's Fujinami?"

He just shook his head at the question.

"He was dead when I got to him."

The pain in the large man's voice quieted all those around them as many realized for the first time that Fujinami was not there.

Somehow, it came as more of a blow than Chowdhury would have believed. She had lost friends in combat many times before. It is never easy, but this loss was especially hard. Jens Fujinami had been the most gentle, alive man that she had ever met. It did not feel real. She thought back to her last sight of him and how incongruous it had been. This quiet, peace-loving man, taking on an armored Marine with a sharpened stick. And taking him down. But apparently losing his life in the process.

The cabin was very quiet as everyone dealt with the loss in their own way.

They were snapped back to their duties by Captain Brighton handing out assignments. While grief filled their hearts, there was still work to be done and returning to previous routines helped many to cope.

They had constructed two manually-pedaled generators to create power in order to operate the ship's systems, like recycling the oxygen and water that had kept us alive. Energy was no longer supplied by the batteries that had been removed by Teach and the other pirates. Everyone took a shift on these generators to keep essential ship systems running, even the captain. Even though they now had been able to collect a small number of battery cells for the ship, they continued to work the generators, saving the batteries against a single, hopeful attempt at a jump point that might allow them to escape their exile in this system and save their lives. The jump was by no means a certainty. They had no chart of the jump point and only the vaguest idea where it lay. They were almost out of food and many of the crew members were already beginning to have their bodies shut down from lack of nutrients.

Mapping the system seemed to take a lot of the captain's attention, but he seemed to relax with a task to concentrate on. Despite the many changes and disruptions of the past few days, the crew quickly slipped back into their old routines.

Captain Brighton continued to come down for his shifts on the generator, even though Chowdhury could tell no one expected him to. He never failed to complete his full shift, though there were others who were struggling to have the energy to finish theirs.

Chowdhury had just finished her shift on the generator and was working her way back to her spot to lie down when Captain Brighton called from the cockpit for Johnson to come forward. Fyonna made her way quickly up the ladder. Once she was in the copilot's seat and Williams was back below,

Captain Brighton immediately called for everyone to prepare for zero g and the rumbling of the thrusters stopped completely.

Chowdhury unstrapped from her place against the bulkhead to move forward and almost ran into Brighton and Johnson as they moved down out of the pilot's compartment into the crowded lower deck.

Brighton looked at all of the haggard crew as they hung in their mismatched straps.

"One of the thrusters is misaligned and we are not able to hold a stable course. Someone will need to go out and make repairs."

"Didn't Long just fix those? Make him go out."

Chowdhury couldn't tell who had yelled and wasn't sure she recognized the voice, but the rumble of the rest of the crew showed that they had general support in their sentiments.

Chowdhury looked at Long and thought about her interrupted 'counseling session' with him. He had been a thorn in the side of Brighton and the rest of the crew since they got on *Vanguard*, and she had hung him upside down from a tree and threatened to kill him if he ever said or did another thing against the Captain. With the safety of the whole crew at stake she might even have had to follow through with her threat if the DaGaman Marines hadn't attacked the camp. As it was, she had cut him loose and told him to run for the ship while she had gone back to help support the evacuation.

Steve Long looked at Chowdhury with a bitter look, but never said a word in his own defense.

Captain Brighton cut through the noise with a quiet declaration.

"It is no one's fault Lieutenant I will go out and take care of the thruster."

The crew didn't take it that well.

Tim O'Neill stood shakily and grabbed a strap to keep from floating in the zero gravity.

"Captain, let me go out and fix it."

The captain shook his head and pushed off into the airlock area without answering. Brighton began to pull out the EVA hardsuit pieces from their stowage bins and let them float until he had them all ready, then he began to don them as O'Neill, Johnson and Chowdhury entered the small alcove with him.

"Captain, I'll go. We can't afford to lose you."

Tim O'Neill continued his argument as the captain continued to don the hardsuit and prepared to go out and repair the thrusters. They hadn't been aloft very long before Lieutenant Johnson had noticed the misalignment problem.

Brighton shook his head again.

"No, O'Neill. You are in no condition to be outside the ship."

Chowdhury admired O'Neill's attitude, but he was the last one who should be going out. Of the three or four people qualified to do the job, she trusted

Brighton the most. Of course, she trusted Brighton without question, always. O'Neill was having trouble just holding his position next to the Captain and helping him dress in the zero g. His face was pale and he was sweating and breathing hard in the cool atmosphere of the cabin. Long looked healthy enough, but she still didn't trust his attitude adjustment. Besides, he needed to rewire the second generator into the ship's systems to enable them to use the airlock. The pumps and other airlock systems would not operate without power. Mackey was another possibility, but while he seemed willing and fairly capable, it was far outside his area of expertise.

"Captain, you're right that O'Neill is not up to it, but he is correct that you are invaluable to the crew. Let me take care of it. Please, sir," Mackey said pleadingly.

Chowdhury shook her head. Mackey could have saved his breath. Once Brighton had decided on the best course of action, he had looked at all of the options and chosen the one with the greatest chance of getting the results that he was after. She listened to the conversation with only half her attention. The rest was focused on the weaponry she had appropriated from Rockhead. She had moved back into 'her spot,' which was directly opposite the airlock, to clear space for others to help Brighton. She had known him long enough to know they would not talk him out of his chosen course of action.

"How much time do you have EV, Bosun?" Brighton asked simply. "I happen to know that you have never been out of a ship in a suit. The Marine unit from the planet may be along at any moment and we don't have the time for you to acquire the necessary skills. I appreciate, as well as any of you, the risk that I am taking, but it is the best solution. There are a few things that you are all unaware of," he said raising his voice and addressing the whole crew. "You know that there is a Marine unit of the DaGama Family operating in this system. They posed as terraformers on A3 and returned to attack us on B3 with the goal of taking *Vanguard*. If they are successful, none of us will get home. Some of you may have guessed their identity already. If they are tracking this ship, then they may be here before we can finish repairs. We must restore the ship as quickly as possible."

Captain completed suiting and waited with helmet in hand while Long made the final connections on generator two and then Brighton turned to the XO.

"Lieutenant Johnson, the plan and courses are recorded in the log. You will proceed as you see fit as commanding officer should anything unfortunate happen while I am out," he said, as if he were leaving his office for a few minutes to run to the store. "Keep in mind that you must not allow this ship to be boarded and *any* of us is expendable to insure that does not happen."

"Aye-aye, sir."

With that, he gave a slight jump and set himself moving for the inner hatch. Alcaraz peddled his generator harder to produce the power necessary to open the inner hatch. Quèneau maintained a steady pace on the other generator to his left.

As the inner door closed and sealed, Johnson held onto a rail with her left hand and pressed the cycle button with her right. She continued to watch as Brighton snapped on his safety line and made his way gingerly aft to examine the thruster units.

The quiet of the cabin was unbroken except for the rhythmic pedaling of the two power sources. Many withering looks were aimed at Long, who seemed either not to notice or not to care, though his gaze turned toward Chowdhury many times. The Captain quickly became lost to Johnson's view and the tenseness in the cabin ratcheted up another notch. Chowdhury continued to make modifications to the rockets within the weapons pack, to give her mind and hands something else to focus on.

And they all waited.

Chowdhury glanced up from her work and looked at the forward chrono. It read 18:50, 4 August.

All eyes in the cabin watched the airlock. If the captain didn't come back, they would all die.

CHAPTER 2

Antoc System
5-8 August 2787

Time crawled as all eyes seemed glued to the chrono in the front of the cabin.

Chowdhury knew the work outside would not go quickly, so she settled in for a long wait. It was a skill she had learned long ago. Eventually, everyone else seemed to relax slightly and get comfortable as well. It was hard to forget that their lives were hanging in the balance but no one can stay on that razor's edge forever.

The inevitable small talk started up. In whispers at first, then gradually louder. Many brutal comments were leveled at Long. As always, Long seemed impervious to any criticism.

Chowdhury remained silent as she continued to work on her munitions. The criticism served no purpose. Johnson apparently agreed, because she called a halt to the comments.

Finally, Chowdhury pushed off from the cabin wall very gently and stopped herself with one of the seat mountings. She retrieved her jacket from where she had tied it onto the bracket. With only one generator on environmental systems, the cabin was starting to chill. She donned her jacket and, hooking a foot through the bracket, continued to wait. Two of the three young ensigns, Hayes and Mitchell, were squirming in their straps under her gaze and finally moved away from her to help with the generators.

The quiet waiting resumed. It seemed like forever, but since the suit only had air for 4.5 hours, they had to be within that limit. She glanced at the chrono again, for the fourth time in the last five minutes, as she had vowed that she would not do. It read 22:03. Brighton only had about seventy minutes of air left. Maybe a little less if he had been exerting himself. She tried to force herself to relax. Brighton wouldn't push the limit, she told myself. Truthfully though, she wasn't convinced. She caught Fyonna

Johnson's attention. Johnson looked worried and tense. Chowdhury thought back to her own worry at her first independent command and gave a slight chuckle to herself. This was truly being thrown into the deep end. Once Johnson was looking in her direction, she said, "With your permission, I would like to suit up and stand by within the hatch in case I'm needed."

"Permission granted."

She was halfway into the suit when the outer hatch started to cycle. Mackey had taken over on the second generator and he also increased his speed slightly when it closed and the pumps engaged and started to equalize the pressure in the airlock to the pressure inside the ship. By the time the hatch was ready to open, Chowdhury had removed and stowed her suit.

The slight burst of air that accompanied the opening of the inner hatch was even colder than the air inside the cabin. Brighton immediately began to strip off the hardsuit as Johnson and Hayes held on to either shoulder to give him some leverage.

As soon as he had his helmet off, he handed it to Williams and said, "Well, it should hold for a while. It wasn't number two after all; it was number four. One of the bolts was missing. We must have had some sort of extra vibration on lift-off. They should all be tight now."

The loudest of Long's critics seemed to look abashed that he hadn't been responsible after all. He had never worked on number four thruster. Long continued to ignore everyone.

"Ensign Mitchell, assist Mackey and Long in restoring the second generator to the environmental systems."

"Aye-aye, sir."

"Everyone else, find a spot on the 'floor'," he said, pointing at the rear bulkhead. I'm going to give everyone five minutes to get settled and then we will burn at 1/8 g for fifteen minutes to put some weight on everything. At that point, we will resume our 1 g burn. Let me know about any issues before that time."

A chorus of accepting affirmations followed.

Chowdhury found a place on the back wall moments before the slight acceleration kicked in. This was meant to let any loose objects fall gently to the bulkhead without becoming missiles. It also allowed everyone to settle gradually instead of being slammed to the deck.

After nearly six days of being on the planet and having room to stretch and move at will, it was exceptionally hard on the crew to be constrained back into a small patch of bulkhead. Tim O'Neill curled himself into a ball and slid as far into the corner as he could get. It seemed that larger individuals were even more careful about violating the personal space of others.

In time, normal weight returned as Brighton went to our standard acceleration. It barely registered to Chowdhury as she fell asleep.

The next morning, it was obvious that many of the crew were in bad shape. O'Neill had not moved in his corner and could barely be roused for the morning 'meal.' George, Ward and Quèneau were not much better. All four were content to lay in their chosen spots and sleep.

Chowdhury felt a pang of fear. She had hoped that everybody would gain more strength from their days on the planet. The more plentiful food and the ability to move and exercise muscles had helped her a lot. However, after only about three days back aboard ship, bodies were starting to shut down again.

Brighton came down the ladder with their morning rations. Many looked up, but most were too absorbed in their agonies to pay attention.

"I have an announcement," he said in a voice full of strain.

"After a review of our rations and flight plan, and in view of the physical condition of many of you, I have decided that we will triple the ration for the next two meals."

Chowdhury was glad to hear this news. Her strength had just started returning after their long ordeal leading up to the planet B3, and now after three days on short rations, she had lost nearly all of her gain. Most were excited about the news, but she also saw the gamble that Captain was taking. He was betting everything that *Vanguard* would be able to make the jump at the jump point. If he couldn't pull that off, there wasn't enough food left for any other options. The captain was throwing the dice on a single plan with no fallback position, something Chowdhury knew was anathema to him. She felt a chill that had nothing to do with the overworked environmental system.

Most of the others didn't know Brighton as she did. She had known and worked with him off and on for nearly twenty years and one of the earliest lessons he had tried to teach her was that you *always* had to have a fallback plan.

Everyone eagerly took their share and ate. Mackey and Long were taking their shift on the generators. Everyone had shifts more often now to accommodate those who could not manage their share. The shifts had been shortened to thirty minutes because that was all that most could perform. Mackey, Long, Hayes, Roberts, Mitchell and Chowdhury were taking the majority of the turns, because they enjoyed the best health of the group. Brighton still took his usual turn but he had taken to staying more and more in the copilot's chair, even when not on duty. Even his strength was beginning to run out and he didn't wish for the crew to see him struggle. The dinner meal came eventually, and the tedium of the day continued. Johnson asked Chowdhury to carry the shares to the cockpit for Williams, who had the duty, and Brighton, who had not come down yet. She took the six small portions and climbed the ladder to the pilot's area.

That task done, she returned to her area in the corner farthest from the generators. She watched O'Neill holding his rations loosely, too weak to eat them. Quèneau and Ward weren't much better.

"Hayes, help Quèneau with her dinner. Mitchell, you help Ward. Roberts, you take Kara," Chowdhury said. The ensigns all jumped up as if she had goosed them.

Chowdhury moved over to help O'Neill. She broke his rations into smaller pieces and placed one on his tongue. That didn't seem to help either.

"Johnson, could you get me some water, please."

When it arrived, she broke his rations into it and got them to dissolve. Working little by little, with his head in her lap, she got him to drink a little over half the small cup. After that, he wasn't able to drink any more.

She laid him down with her jacket for a pillow and returned to her corner.

"Five minutes to turnover," the captain yelled. Everyone got busy strapping in as best they could.

Chowdhury tended to O'Neill and then strapped herself down.

The weightlessness, movement, weight sequence of turnover was never fun. In the shape they were in now, Chowdhury just hoped she wasn't wearing someone else's dinner when it was over.

The rest of the journey was uneventful until the ship neared the area of the jump point. Most of the crew slept as much as possible. Long sat sullenly in the corner as far as he could get from Chowdhury. She smiled when she noticed it. That was fine, she didn't need him as a friend, she only needed him to stay out of the way and let Brighton get them home. She sat in the corner near O'Neill and worked on her gift for Rockhead. It kept her hands busy and her mind off of other things. She looked at the package in her hands and was of two minds on the configuration that she wanted. Finally, she made her decision and went back to work.

There was still a possibility that the group would have to fight their way to the JP. If that was the case, she had made the wrong choice. Of course, in the shape everyone was in, if it came to another fight, they were finished anyway.

As they neared the jump point, she had Long reconnect generator two to the forward airlock and had moved her bundle inside.

When Brighton announced twenty seconds to jump, Chowdhury climbed the ladder and vented the airlock to space. The improvised bundle tumbled out and continued away. She was happy to see that there was no ship here to block their exit. She had made the right choice with her package configuration and the first ship to come after them would get a rude surprise.

CHAPTER 3
Gerrix System; Gemmill
8 August 2787

Vanguard slid out of the transit field, blue streaks of energy dissipating as its inertia carried it farther into the Gerrix System. Captain William J. Brighton felt the weight of his responsibilities dissipating at the same time. He had kept the vow he had made to get his people out of harm's way, and he felt the blessed relief at having completed that Herculean effort.

He listened as excitement reigned in the cabin. Everyone was starting to realize that the most difficult portion of their journey had finally come to an end. As he looked back, he could see several faces with tears running unabashedly down their cheeks. He took a deep breath and started checking off points on his mental checklist.

Looking at the battery indicator he checked the remaining charge and realized they had more power left over than he thought they would. He took a deep breath to clear his mind of the last of the fogginess from the jump and began to make assignments.

"Alcaraz, we need the long-range communications reconnected."

"Ensign Roberts," he called, "take Mr. Long and reconnect the Gravitas drive."

"Ensign Hayes, discontinue cranking the generators and reconnect all functions back to battery power. We seem to have enough power left for all of our needs. Once that is completed, see if you can assist Alcaraz with the LR-comm unit."

This pronouncement was met with a renewed buzzing of excitement throughout the cabin as everyone finally began to believe their ordeal was, in fact, finally nearing an end.

The officers and crew picked their way stiffly through the packed bodies as others made way for them or moved to help as needed. For the first time in nearly a month, the crew was animated with their hope for success.

Still, with all of their renewed hope, their faces showed some of the uncertainty of the welcome that awaited them. Captain Brighton was still struggling with his decision to come here at all.

Major Chowdhury looked at her old friend and moved slowly up to the co-pilot's chair.

"What's wrong?"

He shook his head and continued to look forward, at the view screen.

"It is possible that I have made the wrong choice in bringing us here."

"Our only other choice was to land on one of the planets and probably die there."

"Considering our orders, that might have been the best choice. We are exposing *Vanguard* and her engines to Portales scrutiny."

Chowdhury was silent for a long time. Brighton didn't force the issue. As his head of security, he was sure this was not the first time she had thought of these issues.

"I still believe this is the right course of action. The engines would have been lost to Warner R & D as well if we had remained in the Antoc system. I have discussed with Lieutenant Johnson the need to ensure the security of the ship. We have no reason to believe the treaties will not hold. We just need to make sure that they have no reason to suspect there is anything of interest on this ship."

He nodded, still not completely convinced. She didn't tell him that she had saved one of the rockets to create a demolition charge to destroy the ship if the Portales officials tried to take it away from them after all. That was her job. Captain would understand and approve if this final solution had to be used. If it didn't, he probably didn't want to contemplate the explosives sitting under their feet.

While treaties had been in place for decades between the Warner Family and the Portales Combine, who controlled this system, Brighton was still unsure of their welcome. Portales had been a firm ally of the Warner Family for many years, but things had grown strained over the last decade. All the treaties were still in force, but they had been increasingly competing for the same resources and planets. Understanding and help might be in short supply. Brighton took several moments to delete any relevant files from the computers but there would be no way to hide the altered engines if they became curious. Chowdhury nodded as she watched these extra precautions. She got up and moved back to her previous position, giving up the co-pilot chair to Johnson.

With the more efficient Gravitas in-system engines running, they had the communication time lag down to about seven minutes as they continued to drive toward the only habitable planet in the system.

Lieutenant Johnson had been able to contact the military base on Gemmill and conveyed enough of their situation to get a response from Medical. Doctor Ward had joined the other officers in the pilot's compartment and detailed the needs of the various crewmen to the medical staff on the surface.

Finally, there was nothing left to do but pilot the ship and relax. Brighton reached up and shut off the receiver. He needed to think without interruptions, to go through contingencies and make sure they had not forgotten any visual evidence that could betray their secret. He switched seats with Johnson, letting her have the honor of piloting the ship for this final leg of their journey. She had learned and grown over the last several weeks. She had earned the right to sit in the pilot's chair.

As they finally got ready to begin their descent to the surface, Brighton turned the receiver back on and sat back into his chair as Lieutenant Johnson expertly piloted the small craft toward the planet. Sweat was rolling down Brighton's gaunt face and soaking into his uniform tunic as he watched Lieutenant Johnson pilot *Vanguard* toward their planetfall on Gemmill. The ever-present chill of the compartment had been replaced by a more normal setting as the environmental systems came back online now that they were again on battery power. They were no longer limited to the reduced levels that had been all their diminishing physical endeavors could maintain with the improvised, manually operated, generators. The additional heat was a welcome change and would improve the overall comfort in the long run, but their bodies had not had time, yet, to adjust to the change. Brighton also hoped that the full environmental unit could do something with the growing miasma created by seventeen sweating bodies in close quarters, but he was not, personally, very hopeful in that regard.

Brighton reached forward to take control of the ship as they began to enter the atmosphere. He noticed a small shaking in his hand as he reached for the control stick. He quickly made a fist to hide the quiver from Johnson and resumed his reach. The shaking was still there so he pulled his hand back to his lap.

Johnson turned to him quizzically.

"The ship is yours, Lieutenant, Take us down."

Her face lit up with pride.

"Yes, sir!"

She hadn't seen his weakness and he was not going to let her know that there was any reason other than her own ability that earned her the tight to land the ship. She deserved the honor, regardless of the reason for his decision.

Their planetfall, when it came, was anti-climactic. After numerous unpowered descents, the computer controlled, anti-grav descent was ridiculously easy.

She showed that his faith had not been misplaced by setting them down without the slightest bump; directly in the center of the landing platform at Gemmill's main spaceport.

Medical needs came first. Captain Farel of the Portales Combine Navy was on hand with what appeared to be the entire medical detachment of the naval base when the hatch opened on the side of *Vanguard*. The hot, dry, desert air swirled into their home and began to replace the overused atmosphere within. The heat was another challenge to the newly adapted bodies. Ricardo Smith, who had been leaning on the wall near the hatch began to wobble and would have gone down if not for the quick reaction of Ensign Hayes, who was standing next to him. Hayes caught him around the shoulders and gently laid him on the deck at his feet. His body effectively blocked the exit, but such was the respect for the diminutive cook that no one complained nor moved. Dr. Ward hobbled to his side as quickly as his own injuries allowed, and the Portales medical team moved in from outside. Ward pushed their helping hands aside in one of the only exhibitions of pique that the cheerful doctor had exhibited during the long ordeal. He finally allowed their help when he had satisfied himself that there was nothing life-threatening involved.

Brighton worked his way out of the pilot's compartment and noticed the still form of Timothy O'Neill. His shrunken frame was curled up in the corner and he had not moved in the general milling toward the exit.

"Dr. Ward," the captain said quietly, not wanting to upset the crew. "Could you check on O'Neill?" Ward never moved. He had not heard the soft question, but Major Chowdhury had. She turned and saw the once burly electronics technician and let out a muffled moan. She moved hesitantly to his side as if afraid of what she would find. She reached down and put a finger to the side of his neck as the Portales crew lifted Smith onto an anti-grav stretcher and moved him to the waiting ambulance. Ward followed him out but no one else moved through the now open hatch. They were all waiting for Major Chowdhury's pronouncement.

"He's alive, sir. I can feel a faint pulse, but his chest is barely moving. We need to get him out of here to where help is available," Chowdhury said.

Hayes reacted most quickly, again. Finding hidden reservoirs of strength, he jumped out of the hatch and grabbed one of the medical personnel, a petite young doctor who couldn't be more than a year out of med school, and physically propelled her into the hatch. The startled medic took one look at the patient and began to work as if she were used to being thrown around. Seeing this healthy, active young medic, made Brighton reevaluate the overall condition of his crew. They were truly a sorry lot indeed. Starvation was

18

evident on some faces and the marks of prolonged depravation on all of the rest. All of these effects were magnified now that he had healthy bodies to compare them to.

Brighton suddenly felt an overwhelming, crushing sorrow. Wrapped around that sorrow was the burning fury that demanded justice for the wrongs his people had been made to endure. He looked out of the hatch to where Hayes had sat down on the dusty concrete and was crying uncontrollably. He was receiving several disapproving glances from the Portales personnel until Mackey, Mitchell and Alcaraz moved between them, creating a wall to shield him from prying eyes. They were all still protecting each other, he realized, even when they could not even protect themselves. He closed his eyes on the sudden sting that threatened to overwhelm him.

He opened his eyes to find that he was alone on the ship except for Lieutenant Johnson and the petite young doctor, Dr. Aruch.

"They are all on their way, sir," Lieutenant Johnson said.

"Where have they gone, Lieutenant?" he asked, trying to get his emotions under control.

The doctor moved forward and bent down to where the captain had seated himself and said, "They will all have at least a short stay in the infirmary, Captain. There is not enough space there for them to remain, however, so the most able will be moved out to other accommodations. Captain Allen, the System Governor, has authorized everyone to stay at his home as long as necessary."

As she spoke, she took Brighton's arm in her hand and was slowly lifting him to a standing position. It was all the captain could do to move himself out of the hatch and into the waiting ground vehicle. He struggled slightly against the pull of her grip, as if he had a reluctance to leave *Vanguard* unattended. Brighton stopped just outside the hatch and ignored the gentle, insistent tug on his arm as if he had some unfinished task to complete but could not remember what it might be.

After Johnson had stepped out into the bright sun behind her captain, she reached up, released the small access panel on the side and keyed the hatch closed and secured it with the keypad.

"*Vanguard* is secured, sir," she said in a firm voice. When he saw Chowdhury nod behind her, he knew it was safe to go.

CHAPTER 4
Gerrix System; Gemmill
8 August

Captain Brighton woke to the brilliant red light streaming through the floor to ceiling windows at the rear of the cavernous room. He had a moment of disorientation before the events of the previous weeks came crashing back to him with full force. He again felt the crushing weight of losing his ship and wondered that he could have forgotten that devastation, even for an instant. He sat up in his bed to take his bearings.

It appeared that the majority of the crew had been brought to this infirmary. It was one large, communal, room with the beds arranged in rows. It had curtains to segregate for privacy as needed. He checked his arm where he was connected to several tubes and monitors. That would not do. He needed to be free to move about the base to secure a way for them to get back to Earth. And quickly, before Teach and the others could make good their escape.

He disconnected himself from the tubes and monitors. It was evidence of the chaos his crew had caused that no one came to check on the loss of signal from those monitors. He found that he was still wearing his uniform pants under the blankets and a short search found the rest of his uniform. Once he was dressed, he started to pace and to think.

He had been so exhausted by their journey from Antoc that he had not been able to properly plan beyond this point. He realized that several things needed to be handled quickly. Just as he could not allow *Vanguard* to be taken by the DaGaman Marines that had attacked them in the Antoc system, he could not allow her to be confiscated by the friendly forces here in Gerrix either. So far, they had taken her to be an unremarkable ship's launch, and paid little attention to her, but he needed to get the ship off of this planet quickly, before they could discover her secrets. He needed to carry word to Earth and Gateway of the loss of *Pathfinder* as well. There were parts of that message that could not be trusted to a communication either. Complicating that duty was the fact that many of his crew were not ready to move and probably would not be for some time. As he pondered these facts, he heard a familiar voice raised in agitation.

"Let me up! I'm not an invalid."

. Turning to look in that direction, he saw Doctor Leonard Ward trying to rise from his bed; to the consternation of several medical personnel.

"Doctor Ward. You will lie down in this bed or I will sedate you and strap you down! Is that clear?" said a young doctor with a forcefulness at odds with her diminutive stature.

Suppressing a smile, Brighton moved to help the medical personnel.

"Doctor Ward," Brighton called softly as he approached, "is there some reason you are unable to lie down properly?"

"What?" the irascible doctor asked as he turned to see his captain approach.

"Is there something about your bed that needs repair?"

Brighton changed the question, enjoying the consternation on Doctor Ward's face.

"No, I was just–"

"Wonderful. Then you can lie still and free up Doctor Aruch to show me around the facility and to visit our wounded."

"Yes, sir," he said with resignation.

"I wish I knew that trick," Doctor Aruch said under her breath.

Captain Brighton took no notice of her muttered comment but motioned to the aisle for her to precede him toward the other patients.

"Captain, you should not be up and about, either," she said with forced severity.

"Alas, Lieutenant, I have duties that will not keep, and after my nap I feel quite refreshed. I do thank you for your concern. Tell me," he continued without waiting for a response, "how is O'Neill doing?"

She looked at him for a long moment as if deciding whether to allow herself to be distracted, before finally answering.

"He is still very weak. We have him on fluids and have put him into a medical coma to aid his healing."

"I wish to thank you for the excellent care you are giving my crew. I need to see the governor this afternoon, but if there is anything I can do to aid you, please let me know."

"Captain, the thing you could do to aid me the most would include getting back into your bed and postponing any visits to the governor or anyone else."

"Alas, Lieutenant, as I said, I have duties that will not keep. I will return as soon as I have had a chance to talk to Governor Allen."

* * * * *

Brighton sat back as he finished his tale and watched the disbelief cover the Governor's face. The warrants that would cause all Portales ships to join in the hunt for *Pathfinder* sat on the desk between them, but the Governor made no move to sign them.

"Surely, you are not serious," he said finally.

The beleaguered Captain stared back at him for several moments. He had met Admiral Allen many years before, when the governor had been an officer in Combined Fleet. Brighton had been a newly-promoted lieutenant on *Resolve*, the flagship of the task force which the Admiral had commanded. Brighton had been gratified to hear that he was now serving as the governor on this Portales-controlled world. It made things much less complicated to deal with someone who knew you.

Brighton had always been favorably impressed with the admiral and always admired his assurance, the ease with which he commanded the respect of his subordinates. Rather than the fish which he was currently imitating, the admiral had always reminded Brighton of a bull; an irresistible force. Physically, he was quite imposing. He towered over all his crew and was, indeed, even taller than Brighton himself. But his presence was more than mere size. At well over a century in age, his charisma should have diminished with the decline of physical energy, but the opposite seemed to be the case. The silver hair and creases on his face simply multiplied the effect that he had always had on his crew.

Brighton steeled himself against the admiral's response and charged ahead.

"I assure you, sir, I have never been more serious in my life. If you would be so kind as to sign the warrants to hold *Pathfinder* if she should show up in Portales space and forward them to your factors throughout the frontier, then we will be out of your hair as soon as my crew has sufficiently recovered."

He sat and stared at Brighton, spinning his pen idly in his hands. Brighton sat perfectly still by effort of will. He could not be completely honest with this man who represented a competing government. The secret he held of the true nature of *Pathfinder* and *Vanguard* was too valuable and too important to the future of the Warner Family.

"You cannot be remotely ready to go. You and your crew spent 46 days traveling over 23 billion kilometers with only reaction thrusters. The feat is unheard of. If I hadn't heard the tale independently from your Lieutenant Johnson and the rest of your crew, I would have stated that it could not be done. You all need to recover your strength. You have five cases in the infirmary that Doctor Aruch has told me she is not sure will survive. You have already done the impossible. Don't push your luck."

"I quite agree, Governor. The feat is clearly impossible when looked at in its entirety. Nevertheless, it had to be done, so we did it. My crewmen are quite remarkable in what they are able to accomplish. They will be able to complete the journey."

"I have no doubt that they could accomplish whatever they set their minds to, but that is not the issue," he said as he sat back in his chair and tossed his stylus onto the stack of hardcopy warrants. "I'll tell you what I will do. You agree to spend seven days here for your crew to recover and I will arrange the credit you have requested. I will also supply you with a ship, on loan, of course. I have several *Wayfarer*-class transports that are semi-mothballed here at Lazarus Station. It will take at least six days to prepare one and certify that it is space-worthy. Even with that delay, you could be back in Earth space in twelve days. That is much sooner than your planned route. The *Wayfarer* class can tow your *Vanguard* or you can leave it here to be recovered at a later date, whichever you prefer. She might even be able to fit inside the main hold. As you know, the *Wayfarers* need a crew of 45, so I will arrange for additional crew to help out with the return trip so that you don't all die on this fool's errand."

Brighton sat there with his head spinning. The governor had just offered nearly everything that was on his "would be nice, but not likely" list. Yet twelve days was entirely too long. They must return to Antoc before *Pathfinder* could get underway. If Teach were allowed the time, he would make good his escape. Allen was correct, however. Twelve days to Earth was better than they could do on their own. Most of the relief, however, was due to the fact that the governor did not suspect the true nature of *Vanguard*. If he had, he would not let it out of his sight, no matter what protests Brighton made.

"Your Excellency, I accept your offer with many thanks," Brighton began quietly. "I'm sure that we will not need any extra crew, however. My crew is up to the challenge of manning a functioning ship with complete food stocks. As a matter of fact, they might consider it a vacation. I ask for one modification, if I could?" Brighton said, trying to make it seem as natural as possible.

"What is that, Captain?"

"Sir, if it would be possible, could we get the batteries replaced in *Vanguard*? I would like to use it to go on ahead while my crew recovers, so that I can begin putting together our response to the theft. I will leave Lieutenant Johnson here to take charge of the refit and the injured crew."

The governor sat back in his chair once again, pen still spinning through his fingers, and took in the measure of the man in front of him. Brighton sat very still and tried to act as naturally as he knew how.

"Brighton, when we were together on *Resolve*, I wasn't sure you had what it takes to command. You seemed, at the time, to be too rule-bound and rigid to lead a crew. I am glad to see that I was wrong in that assessment. If that is what you feel you must do, then I will support you."

"Thank you, sir."

CHAPTER 5
Warner Station, Earth Orbit
17 August

"Admirals, with all due respect, I feel that your decision is the wrong one," Captain William Brighton said with a calmness that covered his rising fury. "*Pathfinder* is still there to be collected, if we hurry."

Brighton had endured interviews and questions ever since he had returned to Earth, and he knew the added stress of dealing with the press made him short-tempered. He was doing his best to sound calm and collected.

"You don't know that," retorted Vice Admiral Chohiro Tominaga. The diminutive admiral made no attempt to hide his inner emotions. The vitriol in his comments was unmistakable. "We would not be in this position if your judgment were reliable. Why should we compound your mistake?"

"We need to reclaim—" Captain Brighton began before being cut off by Fleet Admiral George Turley.

"Excuse me, Captain," he said to Brighton before turning to Tominaga. "Regardless of past events, Cho, we cannot let this play out without making the attempt."

Admiral Turley was the physical and emotional antithesis of Admiral Tominaga. Where Tominaga was small and fiery, Turley was above average in height and deliberating in his manner. To this point, he had allowed his aggressive fellow Board member to direct the proceedings of this meeting of the Warner Fleet Governing Board, as had the third and final member of the board, Vice Admiral Karina Kuznetsova.

The three admirals sat behind a thick wooden conference table and looked down at Captain Brighton in his position slightly below them, behind a small desk to their right. The tableau created the impression of a penitent pleading before his betters and had the effect of setting off Brighton's temper even before Admiral Tominaga had gone on the offensive.

"We can't just throw good money after bad, either, George," Tominaga replied, doing a better job at hiding his anger while addressing his peer.

"I'm not sure you are grasping the salient point, Admiral," Turley stated quietly before addressing the large man seated behind Brighton. "Admiral Cosina, have you discussed this issue with Chairman Warner?"

"Yes, sir," said the bear-like admiral as he stood and stepped over to stand next to Captain Brighton.

"Could you please summarize those discussions for us, specifically in regards to the consequences of not recovering the experimental technology that is inside *Pathfinder*?"

"Certainly, sir," Cosina began. "It is the opinion of Chairman Warner that the consequences would be the bankruptcy and dissolution of the Warner Family, sir."

"Preposterous!" shouted Tominaga. "Are you trying to insinuate, Admiral, that if we don't go along with this ridiculous plan, the Warner Family will shrivel up and die? This is the worst job of emotional blackmail I have ever seen. It isn't even believable."

"I can only answer the questions that are asked," Cosina said through clenched teeth. "The opinion in question is not mine, but rather that of Chairman Warner."

"Do you concur with his assessment, Admiral?" Turley asked before Tominaga could interject more invective.

"I do, sir."

"Why do you hold this opinion?"

"Sirs," Cosina moved his head side to side to include Admiral Kuznetsova, "in the last four years, revenues from Gateway Interspatial have amounted to 64% of the Gross Familial Product. In the hands of our competitors, the experimental engine technology aboard the test prototype, *Pathfinder*, would eliminate that revenue. If the revenue could not be replaced, the Family would soon be unable to meet its financial commitments."

"Do you see any other ways to accomplish this mission?"

"No, sir, I do not. I can see several ways to proceed with the development and rollout of this technology in our own regards, but that would allow other Families to have this technology and to bypass our gates and the tariffs that they generate. Those revenues would be partially offset by the licensing fees on the engines themselves, but if we do not pursue and recapture *Pathfinder*, I see no recourse other than that described by our CEO."

"Very well, Admiral, if you and Captain Brighton will excuse us, we will deliberate and let you know our decision. In the case that we decide to go forward with your recommendation, please have a deployment plan ready."

"Yes, sir."

CHAPTER 6
Warner Station, Earth Orbit
19 August

Captain Brighton stood in front of the Warner Space Museum and watched as they towed *Vanguard* into its final resting place. He should feel honored to have the small ship given such a place of honor here. It would always be in commission and it would reside between two of the most revered vessels in Warner Naval History. He wasn't sure the feat was equal to the reward.

In front of *Vanguard* floated *Courser*. The sleek exploration ship made the launch being towed into position seem small and clumsy. Brighton was reminded of the first time he had seen *Courser*. He had been about to graduate from the academy and had applied to Captain Cosina for a position aboard the ship. His achievement of gaining that posting had directly resulted in his crew surviving to return home in *Vanguard*. It had been while serving aboard *Courser* that he had first visited the Antoc system when they had discovered and explored it. His knowledge of the pre-positioned stores that had been placed there for follow-on explorations that had never materialized, had allowed them to scavenge the batteries that had allowed them to escape.

What a small universe it seemed to be sometimes. Of course, the third ship tethered there, Admiral Overman's Cruiser *Fury*, dwarfed them both.

"Captain?"

He was brought back from his thoughts by a tap on his shoulder.

"Yes, Major?"

"Everyone has arrived."

Brighton's face registered a moment of surprise.

"Everyone?"

Chowdhury gave a small snort.

"Well, Delacoeur didn't make it, of course, but everyone else is here. "

Brighton surveyed the group that had collected behind him as he was woolgathering. It did appear that nearly all of the group that had escaped with him from Antoc had assembled at his request. Even Dr. Ward was in attendance, though it looked like he had escaped from the infirmary.

He cleared his throat and paused, reluctant to begin.

"I know that you have only been home for a few days and much of that time has been consumed with interviews and testimony concerning your actions during the piracy that the media is insisting on calling 'The Pathfinder Mutiny.' I am truly sorry for that. Several things have arisen that necessitate an urgent departure. Because of that, I asked for you to assemble so that I could inform you all personally of the situation.

"Firstly, I had the opportunity to speak with Admiral Cosina a few hours ago, and the Board has approved my plan to return to Antoc in pursuit of *Pathfinder*. If we are able to return quickly enough, *Pathfinder* will likely still be in the system. They have given me a flotilla of four ships with which to track down Teach and his fellow pirates. One of my ships is just coming out of refit and needs a new commander. If you are willing to forgo your leave, Lieutenant Commander Johnson, the command is yours."

Johnson jumped.

"Lieutenant Comm... Sir, I'm not due for promotion."

"Nevertheless, the Board was impressed with your actions and authorized the promotion and the ship. What is your answer? Are you willing to bypass your accrued leave?"

"Yes, sir."

"I had hoped that would be your answer, so with your new rank, Lieutenant Commander Johnson, you will take command of the destroyer *Yargus* and accompany me back to Antoc. I have convinced Fleet that *Pathfinder* has no way to leave that system quickly and if we can respond before they make repairs, then we will be able to recover the ship."

Her face showed some concern.

"Will you be returning on another ship?"

Brighton nodded to the group in general.

"I have been given overall command of the squadron consisting of *Yargus* and the light cruiser *Dagger*. Due to the nature of your last experience in the Antoc system, you are all due as much leave time as you feel necessary before returning to duty."

He turned slightly to face Johnson while still including the rest of the group.

"We will depart at 1900 hours tomorrow. Any who wish to accompany us back there to see justice done will need to report for duty aboard *Yargus* by 1600. I have the permission of the Board to accept any of you into the crew who reports aboard by that time. I won't ask anyone to come with us. You have all done enough. I am very proud of all of you."

This last was delivered almost as a whisper.

Dr. Ward stepped forward and there were tears rolling unashamedly down his cheeks.

"I won't be able to come, sir. I'd give anything to sail with you again, but I won't be released in time and well..." he looked back at the young blonde woman on his arm and shrugged. "You understand, don't you, sir?"

"Indeed, Doctor. You have other duties to attend to. We will take care of everything else for you."

The doctor stood as straight as possible considering his bandages and delivered his best parade ground salute. "Thank you, sir," he said simply, but Brighton could tell that he was referring to more than just these words.

Derrick Mackey stepped slightly forward to get the Captain's attention.

"Beg pardon, sir, but would it be possible to accompany you on *Dagger*? Meaning no disrespect to Commander Johnson."

Before Brighton could answer, he noticed Admiral Cosina emerge from the shadows on the opposite end of the corridor. The admiral began to speak as he approached the group.

"That won't be necessary, Master Chief Mackey,"

The group turned to face the oncoming admiral. Most straightened unconsciously in the presence of the large, bearlike legend.

"There have been some changes in the plans since I last had the opportunity to speak with your captain."

He looked to Brighton.

"If I may?"

"By all means, sir."

"Captain Andrus has taken ill aboard *Dagger* and her XO is a Lieutenant not yet ready for command. Consequently, we have had to shuffle commands among the officers available in-system in order to make our departure schedule. As soon as her new captain reports aboard, *Dagger* will be ready to get under way."

"With respect, sir, I don't see how that will change the situation," Mackey replied with a slightly confused scowl on his dour face.

"Oh yes, I forgot to mention that, didn't I? Lieutenant Commander Johnson has been assigned as *Dagger's* new captain, and she will be Commodore Brighton's new flag captain. Lieutenant Ramirez will be promoted to command *Yargus*. Any officers or crewmen that wish to accompany Commodore Brighton will need to report aboard *Dagger* before the appointed time. Major Chowdhury, if you are willing to waive your leave time, you are requested to report aboard to take command of the Marine contingent on *Dagger*. Does that clear things up for you, Master Chief?"

Cosina walked away as the crew started talking excitedly amongst themselves. Brighton watched him go for several moments before detaching himself from the group with a glance at Chowdhury. Receiving a nod in return, he knew she would take care of everything here. With that done, he followed Cosina.

He caught up to him just around the turn of the concourse. Cosina was leaning on the railing looking out over the displays in the gallery below.

"*Dagger* is a light cruiser; *Rapier* class. There will be resentment at her being given this command."

Cosina looked up at Brighton, used to his blunt approach.

"Do you think she's not ready?"

"It is not that. She is ready. But this will make things much more difficult for her within the fleet."

Cosina nodded without looking at Brighton.

"If she's ready, let her have it. The rest won't hurt her."

Brighton stood stiffly looking at *Vanguard* in the distance.

"Vanguard shouldn't be out there with the rest of those ships."

Cosina stood to his full height. He overtopped Brighton by a few centimeters. He stepped in closer to his former protégé.

"Do you know why that ship is there?"

"I just admitted that I did not."

"That ship is there for the same reason that Johnson was promoted out of turn. For the same reason that every member who served aboard her during your escape has received the meritorious service cross. For the same reason that Chowdhury received the Marine Distinguished Service Star - again. Your escape has to be the single greatest feat of astrogation that I have ever heard of. Admiral Graves agreed, and since you refused to accept the awards that were offered, other accolades were found."

"They were totally appropriate for the crew but I deserve nothing until I have been able to bring in Teach and recover *Pathfinder*. I thought I was clear on my position."

Cosina stood straighter and looked at Brighton for several moments before replying.

"You were. You should probably visit each ship and check on their readiness. You will want to be sure to meet your scheduled departure."

With that, he turned and walked away.

With a glance over his shoulder at Chowdhury, Brighton started walking also.

"We might as well start with *Yargus*."

"You could save that one for last."

"No, best to get it over with."

* * * * *

Captain Julio Luis Ramirez-Flores strode into the briefing room with a wide smile on his swarthy face. He was casually moving his shoulders at every opportunity in order to show the lieutenant commander's pins and command tabs that were now attached there. *That was quick work,* Brighton thought. *He was only notified of his promotion and command less than two hours ago.* Brighton watched as the dusky young officer moved down the length of the table to take a chair next to Chowdhury, opposite the one he himself had taken.

Brighton nodded to the young officer as he settled.

"Things have been moving very quickly for the last few days, Commander, so I want to make sure that we understand each other and, most importantly, that you understand what is at stake here."

He waited until he got a confirmation from Ramirez before continuing.

Let's start with introductions. This is Major Chowdhury. She will be in charge of all security for the task force and for the mission. If she gives an order, you can assume it carries my approval. As you know, we have a scheduled departure time of 1900 tomorrow. I am aware that you are just coming out of a refit. I'd like to hear about your unit readiness, and finally any roadblocks that you see that might delay our departure. Commander?"

"Yes, sir. The crew is still reporting in and I have some civilians aboard finishing up the refit, but we will make the schedule. The new jump engines are nearly finished with trials and Fleet has signed off on them, pending our final tests during the outbound leg of our journey. I don't foresee any other roadblocks, sir."

"Very good. Good job with the refit and congratulations on your promotion."

"Thank you, sir."

After several more minutes of idle conversation, Brighton excused himself to visit the other ships.

"Would you like a tour of the ship before you go, sir?"

Brighton rubbed the corner of the table absently.

"I would enjoy that on another occasion, but I am familiar enough with her capabilities."

Chowdhury snorted softly at his back as Brighton stepped out of the chamber.

Ramirez gave Chowdhury a questioning look as she started to follow him out.

"Did I miss something?"

She smiled.

"Do you know what the name of this ship was before she was recommisioned?"

"Yes, ma'am. She was the WNS *Reigna.* They brought her out of mothballs to install the new engines."

"You might want to look up the name of her last commander before she went into the mothball fleet."

With that she was gone.

She caught up to Brighton on the dockside. He was staring back out through the observation blister at the ship. Chowdhury stood beside him, saying nothing.

"It's fitting, in a way; that she come with us to find *Pathfinder.*"

Chowdhury just nodded. She knew what he meant. *Pathfinder* had been built from one of the sister ships of the same class. The *Risea* had been pulled out of mothballs and extensively refitted to become an exploration ship. Her time serving aboard *Pathfinder* had reminded her of *Reigna* and she had only ridden aboard the older ship back to Earth, not served aboard her as Brighton had.

They continued on around the curve of the station to a small system patrol craft and repeated the introduction procedure.

"Lieutenant Dan Carmichael, CO of the *Avram,*" said the short, stocky, red-haired officer as he stood slowly to salute Brighton as the commodore entered his office. The ponderousness seemed to be as much a part of his makeup as the smile was to Ramirez. "We just arrived in the system from frontier patrol in the Poma system. We were due to rotate out for minor refit when we were transferred here instead. Our engine capacity is down to 87% of rated output and the crew is worn down, but they have been together for better than thirty-six months and they work well together. We are currently in the middle of re-supply, so that is our only possible roadblock. We have gotten priority on everything, though, so it looks like we will be ready, sir."

"Your record has been excellent, Captain. I expect that you, and your crew, will continue to perform at the highest level."

When they had completed the offered tour of the small ship, Brighton could see that the young man had not exaggerated. The crew looked tired. But they were still functioning.

They had to traverse the length of the station to meet with their final captain. The ship was so large, it didn't dock. They were required to take a small shuttle out to the factory ship as it hung just outside the station's orbit.

A heavyset, greying officer met them as they disembarked from the shuttle.

"Captain Larsen?"

"Yes, sir. We've been expecting you. The engineering crews are all aboard and ready; Fleet crews are 90% in. We were loaded and scheduled for departure to Minoa tomorrow anyway, so everything is already in place. I don't anticipate any issues with the schedule you transmitted, sir."

"Very well, thank you. Would it be possible for you to join me on *Dagger?*"

"Certainly, sir. When would you like me to be there?"

Brighton turned a slight smile to the older man.

"Would now be convenient?"

He knew that Larsen would not be able to refuse, whether it was convenient or not, but the older man made the best of it.

"Of course, sir."

"And could you have your comm officer contact the other captains and have them meet us there as well? I would like everyone to meet my Flag Captain and have a short briefing."

"Certainly, sir."

* * * * *

Once they had all assembled, Brighton introduced the three captains to Lieutenant Commander Johnson.

"Commander, could you give us a quick update on *Dagger*'s readiness?"

She sat up a little straighter and showed her nervousness with a slight trembling of her hands as she began.

"We are short several crewmen that are still in the infirmary from the outbreak on their last deployment, but Fleet Medical has given the ship a clean bill of health and released us from quarantine. I anticipate a large contingent of replacements before 1600 tomorrow that will enable us to replace the majority of those still in the infirmary. There are no other insurmountable problems. We will be ready to move out on schedule."

"Thank you all," said Brighton as he reclaimed his standing position. His presence dominated the small group. Though still thin from his recent hardships, his fiery red hair shone like a beacon that advertised both his temper and energy. "I need to summarize the events that have transpired to this point, so that all of you will understand what our mission is and why we will proceed in the manner that I outline."

He took a deep breath and seemed to grow even larger as a blazing light came into his eyes and his features became more animated.

"Nearly two months ago, on 28 June of this year, a prototype vessel, *Pathfinder*, was stolen from WSN by members of her crew. During the takeover, the ship's propulsion and computer systems were disabled by me. Members of the crew were set adrift in a survey launch and at least two died as a result of that action. It is possible that more were killed during the takeover as we left injured crew aboard. We believe the ship's propulsion and computer systems were sufficiently damaged that the pirates will not be able to regain control of her without extensive repairs.

"I intend to make a high speed run from here to the Antoc system, where we should still find *Pathfinder*. The pirates will be captured and dealt with, as well as those members of rival Families who have set up illegal bases within that system."

He turned his piercing gaze upon each captain in turn.

"There is nothing more important to the Warner Board than the recapture of that ship. Is that clearly understood? Nothing."

"Yes, sir," came the staggered chorus from his captains.

"Very well, return to your ships and let's make sure we are ready for our 1900 departure. Dismissed."

CHAPTER 7

Dagger
19 August

Chaotic was the only word to describe the scene that greeted CWO2 Steven Long as he made his way into one of the cramped enlisted berthing compartments on *Dagger* with Lieutenant Grant, *Dagger's* executive officer, and the other new replacements just arriving on the ship. The lower deck compartment was roughly five meters long and three meters wide with eight bunked beds in a single row jutting out from the wall to the left of the hatch. Five of the replacements were fellow crewmembers from *Pathfinder* and *Vanguard* and Long had already been living with them under the worst possible conditions for the last two months. They were also the closest thing to family that the crusty warrant officer had left in the world.

The other five enlisted personnel were the normal type of crewmen that were scraped up when Fleet needed men in a hurry. Those between assignments or in transit at Earth, awaiting transport to their next assignment.

The compartment was designed to hold sixteen enlisted personnel. There were, however, nineteen ratings sprawled on bunks or lounging on any convenient flat surface. Long's first thought was that he was glad he didn't have to stay. As a chief warrant officer, he would have separate berthing from the enlisted ranks, as would Master Chief Derrick Mackey, who was now the most senior enlisted crewman on *Dagger* and, as such, would take on the duties of bosun.

The new arrivals entered as a group and descended on those who were already in residence. Many dark looks were directed at the newcomers. With the experience of an unrepentant troublemaker, Long knew there was a storm brewing here. He could feel it in the air.

"Atten-shun!" cried out a voice as Lieutenant Commander Grant entered the compartment.

"As you were," the XO responded automatically. The crew stood at ease at the foot of the bunks as the invading group moved into the room.

"Grab any bunk that isn't claimed," Grant directed as he moved off to the left into the shadow of the first rank of bunks to clear a path for those who were moving into the room. They stood in a group, milling around as if uncertain where to go.

Most of the bunks were still loaded with the gear of those who had been transferred off the ship while at Earth. Those sick crewmen had been in the infirmary at Earthstation when *Dagger* had been ordered back into space, and the speed of their departure had not allowed for the belongings to be forwarded in time. The resultant clutter, and the fact that all of the bunks were made, made it nearly impossible to determine which bunks were unclaimed. The existing occupants made no effort to enlighten their new bunkmates.

"Anyone who doesn't belong here, move out," Grant said in an effort to remove some of the confusion from the situation. In response to his order, eleven ratings filed out the hatch with commiserating glances to their fellows.

"Grab a bunk," Grant repeated to the new arrivals. George, Smith, Alcaraz, Paul, Quèneau, and the other newcomers shouldered their duffels from the deck where they had dropped them and moved out without a backward glance at their remaining companions. They moved as a group and took the five sets of bunks at the farthest end of the compartment. They unceremoniously emptied all of the gear from the mattresses and began to stow their gear into newly emptied footlockers.

"Hey, that's my bunk," called a scrawny petty officer to Alcaraz as he lifted the pile of clothing from the foot of the bed and tossed it into the growing pile against the wall.

"As you were," called Grant as the petty officer moved to reclaim his territory. "Find a new bunk. If you had wanted that one you should have indicated your ownership, Loring."

"Yes, sir," PO Loring responded as he braced to attention. Long caught the look of anger that the young rating allowed to flash across his face as the XO turned his back. *Oh yeah, definitely trouble.*

"Get out of my way," called a gruff bass voice from behind him. Long started to turn to find out what the problem was when he was rudely shoved aside by a duffle wielded by a short fireplug with a steel gray buzz cut and a florid face. Long was not sure if the redness was caused by the master chief's obvious anger or if it was the natural state of the non-com.

"Watch your step, Master Chief," Long responded with deceptive mildness. "If you're not sure where to stow your gear, I could suggest a location."

"Durrant," Grant said as he stepped in to curtail any further escalation, "you'll be in bunk one. I'll expect you to get this mess straightened out."

"Yes, sir," the noncom responded, obviously seeing the officer for the first time. With a perfunctory salute, he moved off and cleared the indicated bunk with the sweep of a massive arm and began to stow his gear while completely ignoring everyone else in the compartment.

"Come along," the XO called to the three remaining newcomers and they moved out of the tense compartment in his wake.

"Master Chief Mackey, you will have the bosun's quarters here," he said, indicating an open hatch on the left which had been closed when they entered the area. With a nod to the officer, Mackey moved off to his berth as soon as it was identified.

"Ensign Leslie," Grant said while turning to the petite, blonde officer, "if you could wait in the wardroom, I'd like to go over duty schedules after I get Mr. Long settled."

"Yes, sir," she responded quickly, "I can find my own way. My last posting was on *Katana*."

"Very well, I'll see you there in twenty minutes."

The XO shook his head as he passed the bosun's hatch. The small office looked like a tornado had come through. Long was positive that the tornado was named Durrant. Papers and books littered the floor and Mackey had begun patiently putting the quarters to rights.

Long shook his head as he moved by the hatch. Yep, he surely was glad he didn't have to stay down here.

* * * * *

There are myriad duties that need to be completed by any ship preparing to get under weigh, and *Dagger* was no exception. The fact that *Dagger* was previously not scheduled to depart for two weeks multiplied those duties tremendously. With only one hour to get himself settled into his new quarters, Long now found himself in engineering overseeing the installation of a replacement weapons relay. As a warrant officer on a technical track, Long was outside the official chain of command. Even the lowliest midshipman outranked him. Long, however, outranked the loftiest of the enlisted personnel.

His friend, Master Chief Mackey, as Bosun, was the senior enlisted crewman on the vessel and, as such, received a slight pay boost and had additional responsibilities. In Long's experience, most bosuns were bullies. As the person responsible to maintain discipline on the vessel, this was a natural tendency. Fear often motivated even the worst individual to behave within the set guidelines of military discipline. Master Chief Mackey was the exception. While certainly large enough at 202 cm and nearly 135 kg, he was an even-tempered, patient leader of men who always seemed to gain the loyalty of those under him. Master Chief Josiah Durrant, however, *was* a bully.

As the former bosun, Durrant had lost his position to Mackey due to a trifling six months' difference in seniority. From the bits and pieces of conversation drifting across the engineering bay, it sounded like he was exceptionally unhappy about the change in circumstance.

"I don't care what they say," one of the petty officers in his group mumbled to his neighbor, "it's bad luck to change captains in the middle of a deployment."

"And she's so young," his mate replied.

"I don't know what the blasted hurry is, either. I was supposed to be in Bangkok tomorrow for two weeks' leave."

"Yeah, nothing can be that bloody important."

"Okay," Long cut in, moving next to the complainers, "less talking and more working. We have to have this relay up and tested before the end of watch."

The workers went back to their tasks with a resentful look in his direction. He glared at them until they finally dropped their gaze. *I guess I have a bit of the bully in me too,* he thought with a silent chuckle. The one thing he would not condone was any disrespect to Lieutenant Johnson – Lieutenant *Commander* Johnson, he amended – especially as the new captain.

"Don't worry about it, Mr. Long," said a crisp soprano voice at his back. "I don't think any of them have any idea what brought us to this point. I, however, have been following the news."

Long turned to find Ensign Loren Leslie standing casually at his shoulder, or more accurately, under his shoulder. She stood a petite 157 cm, but didn't seem to be out of place among the large engineering crewmen. One of the new arrivals, she had been assigned as the assistant engineering officer and was nominally in charge of all the work crews.

"What news is that, Ensign?" Durrant called from across the bay in his normal supercilious tones. "Do you know why these prima donnas were dumped on us?"

"Yes, Master Chief, I do know why they are here. Thank you for asking. Now get back to work," she said as she turned her back on the senior noncom and walked to the far end of the engineering compartment to check on one of the other groups.

Durrant stood staring after her for a moment and then turned to a short, rat-faced chief whose nametag read Giovanni and made a comment about ensigns and their over-inflated views of themselves. Long ignored him, knowing that, after his own, personal ordeal on *Vanguard*, he was in no shape to respond in any physical way, regardless of how much he would like to. Durrant followed that comment with his opinion about how the captain had gotten her new position. That brought gales of laughter to his small group.

Long suddenly felt all of the fatigue that had been plaguing him for the last week since the return of *Vanguard* to Warner space. Even the improved rations aboard the Portales freighter hadn't allowed him to regain his strength or stamina, but he also knew that he needed to nip this abuse in the bud or it could get out of hand. Even knowing that he was not physically up to the exertion, he decided that there was very little choice.

In his twenty-seven years in the Fleet, if there was one thing that Stephen Long had learned, even excelled at, it was how to start a fight.

"Was that you, Master Chief?" he asked, stressing the rank and reminding him that he no longer was to be addressed as Bosun. "Blast, I hate it when I'm wrong."

"What do you mean by that?" the Master Chief asked as he stepped forward belligerently, his chin jutting forward and his fists clenched at his side.

"Oh, nothing, I was just thinking to myself," Long sighed inwardly at how predictable Durrant really was.

"Well, if you're going to do your thinking out loud, then spill it."

"Well, I was sharing a beer with a friend before I caught the shuttle, and he said that you were an idiot. I said that you couldn't be. You couldn't have risen through the ranks if you were a gibbering idiot. I was just ticked that I could be so bloody wrong."

"Attention!" came a piercing soprano voice from a few feet away as Durrant gathered himself to swing. The ranks of crewmen snapped to attention at the command in the voice. To his surprise, even Long reacted before he could think.

"Warrant Officer Long, you will take your team and finish the weapons relay installation and then you will return to your quarters where you will remain until your next duty shift at 0800. Is that clear?"

"Yes, ma'am," he said through gritted teeth.

"Very well," she said as she turned to face Durrant. "Master Chief Durrant, you and your group will return immediately to your bunks and you will remain there until you are sent for by the XO. Am I clear?"

"Ma'am, I didn't…"

"I asked you a direct question, Master Chief. I asked, IS THAT CLEAR?!"

"Yes, ma'am."

"Very well, dismissed," she said as she turned her attention back to the other work teams and pointedly ignored both groups.

Long heard barely audible grumbling coming from the other group as they made their way forward to the exit hatch. He shook his head. This was far from over, he knew. He wished the ensign hadn't interfered. It would have been better to get this over with now. This sort of problem always gets bigger until it gets dealt with. Like gangrene, if you let it go too long, something winds up getting chopped off. If things weren't resolved soon, it was certainly going to end badly for someone.

CHAPTER 8

Dagger
21 August

"Master Chief Derrick Mackey, reporting as ordered."

"Come in, Master Chief," Lieutenant Grant responded as he tossed his stylus on top of the stack of data folios piled on his desk. "Have a seat."

Mackey was momentarily caught off guard as the XO motioned to the couch which occupied the right wall of his office instead of the chair which sat in front of his desk. This was certainly not the normal procedure for an enlisted crewman reporting to the second-in-command of a warship during an active mission. Nevertheless, he took his position at the indicated spot.

Mackey would have been more comfortable sitting at the desk for multiple reasons. Since this was his first meeting, and he knew absolutely nothing about the man conducting the interview, the familiar ground of the way such meetings usually went would have put him more at ease. Aside from that, the sofa was too soft, and too low to the ground. When Derrick eased his considerable frame into a sitting position, he sank down so far his knees nearly reached his shoulders. He might have to ask for help to get back out of it, which would make him even more uncomfortable.

The XO followed and sat facing him on the other end of the sofa, giving no indication whether he was aware of the bosun's discomfort or not.

"Master Chief, how are you settling in as boatswain of the ship?" he asked, using the full title for his new position.

"Fine, sir," he replied, again caught off guard by the informal attitude used by the XO.

"Good. Let me explain why I wanted to have this interview," he started. "How many years do you have in Fleet?"

"Thirty-eight, sir."

"In those years, I imagine that you have been a part of many crews, both good and bad."

"Yes, sir."

"What is your impression of this crew?"

"I really haven't been aboard long enough to form an opinion, sir."

Lieutenant Grant studied his face for several seconds. "I can see why you might be hesitant to commit yourself, Master Chief, but let me tell you my impressions and you can make any comments or responses you feel necessary," the XO looked inquiringly at the massive enlisted man until the bosun was forced to nod.

"Good," the XO continued. "The newcomers, mostly from *Vanguard*, have only been aboard for a little more than a day and already there have been three altercations and one incident requiring a Captain's Mast. The old hands were expecting leave and are taking out their frustration at not getting it on the newcomers. The newcomers, on the other hand, have been through a horrible ordeal and are on edge and have no patience, so they are not willing to put up with ribbing that normally they would take in stride. Do you agree so far?" he asked.

"That seems to be a fair assessment, sir," he replied tentatively.

"Very well, Bosun," he said sharply. "That is not an acceptable state of affairs. I have read the entirety of the reports on the *Pathfinder* incident, as well as the Board of Inquiry's reports. Most onboard this ship may not have grasped the situation yet, Master Chief, but I have. We are now jumping to Meridian. Within a few days we will be into the Antoc system, where we will almost certainly encounter hostile forces. By that point, we need to be able to fight this ship as a united crew. It is my responsibility, and yours, to make sure that happens. Am I clear on this point, Master Chief?"

"Yes, sir," Mackey replied, sitting up a little straighter in response to the changing tone of the XO's voice. The new position actually allowed him to unkink his legs a little. He would have adopted it sooner, but he had been trying to reflect the XO's earlier posture of ease.

Lieutenant Grant was the physical antithesis of the prototypical perfect officer. Short and overweight, he would never be posing for any recruiting posters. The large smile that he perpetually wore gave the impression that there was very little thought going on behind his brown eyes, but Mackey had just discovered the opposite to be true. The XO had obviously risen to his current position on ability and intelligence. He had read the reports and understood the effects and consequences of the actions taken. Some of the admirals on the Board had not been able to make that leap. Having lived through the harrowing weeks and the shocks that had been a part of that flight from exile to freedom, Mackey knew at a gut level the stakes they were playing for. Apparently, so did the XO.

"Good," Grant continued. "We only have a few days, so we need to force the issue a little. Here is what I think needs to happen…"

From there, he began to outline his plans. He would take care of the drills to prepare the crew for battle and Mackey would be responsible to work from within and weld the cohesiveness among the crew. He had specific suggestions on tactics for individuals and also had a list of those he anticipated would be the most difficult to bring around.

Nearly an hour later, the bosun found himself walking down the central corridor of main deck toward enlisted country. Derrick Mackey was only a few steps from the XO's office when he heard raised voices from the tactical room. Mindful of the XO's injunction to smooth the two crews into a single whole, he slowed and listened before interfering.

"I'm sorry, ma'am, but that doesn't make sense," he heard Senior Chief Petty Officer Quèneau say in a loud, carrying voice. "That is not the Way It Should Be Done."

"Senior Chief," Mackey heard from across the compartment. The voice was not one he was familiar with but probably belonged to Lieutenant Kim Haskins, the officer in charge of Tactical. "While I am interested in the opinions of everyone in my section, I will run that section as I see fit," she said defensively.

"Ma'am, I understand that the section is your responsibility, but I have a responsibility to point out errors and better ways of doing things."

Mackey knew that Quèneau had been in charge of the Cartography section aboard *Pathfinder* and had run an efficient section. Obviously, this was more than mere settling in. Quèneau should know better than this. Any complaints or comments should take place in private, not in front of the whole section. Because the confrontation was public, the reprimand needed to be public as well. Taking a deep breath, he stepped into the compartment.

"Senior Chief Quèneau," he began in his firmest command voice. "You are out of line. You may make suggestions as you feel necessary, but you should never try to impose your view on the officer in charge. You should apologize now and get back to work."

"Derrick," she began, using the familiarity all those aboard *Vanguard* had come to share, "the reason..."

"There can be no reason for insubordination, Senior Chief," he interrupted firmly. "You know that, now apologize and follow your orders." He turned to Lieutenant Haskins. "Lieutenant, if you would care to prefer charges for insubordination, I will support your decision, but I believe I can take care of this with Ship's Discipline."

"Wha... No, that will not be necessary. Ship's Discipline is fine, Master Chief."

He turned his attention to the red-faced NCO. She looked as if her brother had shot her, confused and disbelieving. "Senior Chief Quèneau, report to my office at the end of shift."

"Aye-aye, Master Chief," she managed to get out.

41

He marched out of the silent compartment without looking to either side and continued to his quarters.

Seven hours later there was a knock on the hatch to his office.

"Enter," he called as he stowed his work.

"Senior Chief Petty Officer Clémence Quèneau, reporting to the bosun as ordered."

"Stand easy. Clémence, this is not the easiest thing that I have ever had to do, so I'm going to go a different direction than you might have expected."

Quèneau made no move to acknowledge having heard him. She was still standing stiffly at attention, her eyes locked on an empty patch of office wall. When she made no response, he continued. "You know as well as I do that what you did was wrong and prejudicial to good order and discipline on this ship. Furthermore, Lieutenant Haskins would have been within her rights to send you to a Captain's Mast, or even a court martial, for your disrespect."

"Permission to speak freely, Bosun," she asked without looking at him.

For a few moments, he said nothing, as if he were trying to decide something.

"Clémence, have a seat," he said, motioning to the chair against the wall in front of his desk.

She hesitated for a heartbeat, then seated herself in the indicated chair.

Mackey waited until she was settled and then continued, "Permission granted. I want this to be a discussion; but before you say anything, I know why you were questioning her methods. You ran your section very well on *Pathfinder*, but you have to let her run her section in the way that works for her. With your promotion to Senior Chief you are the third ranking NCO on this ship. You need to set a better example to the rest of the crew."

"I understand that, Derrick, but it is such a waste," she said with a vehemence that surprised him.

"Clémence, how long have you been in Fleet?" he asked.

"Twenty-seven years. You know that, you were on *Glory* for my first deployment."

"Yes, I was. Do you remember Ensign Tominaka?"

"Yes," she answered hesitantly, thrown off by the apparent change of topic.

"You may not remember, but he was even more laid back than Lieutenant Haskins. I worked for him in Damage Control and he used to drive me crazy. There was no organization or focus. However, when the drive engines blew as we were coming into Gregor, his response was flawless. He saved a lot of lives, including yours if I remember correctly, because his team worked when it needed to. I think if you leave Lieutenant Haskins to run her team, you'll see the same results."

"I'll do my best," she said, though her face still evinced a measure of disbelief.

"You will do sixteen hours of extra duty for the insubordination. Next time, remember to be more tactful."

"Aye, aye, Bosun," she said in clipped tones as she stood and came to attention in front of his desk.

"Dismissed," he called.

As she turned to leave, he added, "You know, if things go well, you'll be back in Cartography on *Pathfinder* before the end of the month. Surely you can last that long."

She didn't respond.

CHAPTER 9
Dagger
23 August

Lieutenant Shannara Burkhalter scanned the wreckage of her engine room and fought down her anger.

She had told them.

This was never a good idea.

The old engines had worked on this ship for nearly a century. She had kept them running for the last six years without any downchecks and they had to go and screw with her systems.

Now two of her people we dead and nine others had received flash burns and would be in the medbay for the foreseeable future. She had told them to leave her engines alone. But they had a 'Better Idea'. She had finally gotten a chance at the Chief Engineer spot and now she was looking bad in front of everybody. She had kept the engines running and that preening idiot Ramirez had taken all the credit. Now he was the Captain and she was going to look like she couldn't keep the engines running without him. The universe was turning against her and there was nothing she could do to save herself or anyone else on this cursed ship.

She took one last look around the scorched and buckled metal and glanced at her chrono. Her five minute venting party was over. Time to get back to work.

"Wilson, is there any usable wiring left at all?"

The slender WO stuck his head up out the bottom of the engine well.

"No, ma'am. We're going to have to put in new runs from scratch. It looks like the capacitor is burned beyond repair also. I've got Hugo pulling it out."

"Okay, I've got Echevarria building a new wiring harness. You get everything cleaned out down there. You have an hour."

"Yes, ma'am."

She turned away from him as his head disappeared back down into the hole under the engine. She left him to manage his crew.

"Jackson?"

"Yes, ma'am?"

How much longer to get this control panel remounted?"

The burley petty officer looked up from the console.

"We should be done in the next twenty minutes, ma'am. It won't be pretty, but it'll work."

"That's all we need."

"We have twelve more hours until we get to the next jump point. These engine are going to work perfectly next time or you are all going to have to walk home."

* * * * *

Lieutenant Commander Fyonna Johnson looked around the bridge of her new command as they prepared to make the final transit in their hectic rush back to Antoc. She had officially taken command only four days earlier and she was still adjusting herself to the new surroundings. More than anything else, she was adjusting herself to her new position as Captain of the light cruiser *Dagger*.

Two months before, she had been content and happy as the helm officer on *Pathfinder*. During the nightmare escape from the Antoc system, she had felt her way through her new position as Brighton's executive officer and all the duties that had entailed; even though they were limited in scope by the situation. Now she had been promoted and given command of *Dagger*.

A newly promoted captain, in her first chance at independent command, usually received a system patrol ship or a corvette. Occasionally, they would get a destroyer, but never a light cruiser; and yet, here she was, commanding *Dagger*, a ship only five years out of the shipyards.

She felt like a player in a drama. *Playing the role of 'the young captain'*, she thought mockingly. Her only consolation, which was not much when you really thought about it, was that her XO, Lieutenant James Grant, was even younger than she and so was not constantly looking at her as if he had a better idea.

Without a doubt, she knew the right things to do, but she felt as if she would be uncovered at any moment as an imposter. Things would have been so much simpler if Brighton had agreed to command *Dagger*. He had simply refused, saying, "I have a ship. When *Pathfinder* is recovered, I will need to be free to resume command there."

She sat back in her chair on the bridge of her new command and surveyed her crew. *Dagger* had a more traditional bridge arrangement than *Pathfinder*. Without the need to add extra engineering, survey, and cartography consoles, the layout was the same as most other WSN ships and so was actually more familiar than *Pathfinder*'s had been.

Lieutenant Russell Bartmist was to her left, at the Astrogation console. Unlike the mercurial temperament of his *Pathfinder* counterpart, he was solid. Though still young, he radiated confidence and steadfastness that was equal to any task. Ensign Amanda Tory had the helm directly in front of the captain's chair as they made ready to transit the last jump point in their personal march back to Antoc.

Commodore Brighton was pacing the deck to her right. He had a station there, with screens that mirrored the consoles around the bridge, but he had too much nervous energy to sit in front of them. She glanced at him and he stopped momentarily and gave her a wry smile; meant to indicate that he couldn't help himself, no doubt. He had been the same at each of the five previous jumps.

At each of the previous JPs they had followed the same plan, *Avram* went through first, the rest of the group held steady and then *Yargus* would follow after five minutes. *Foundation*, the massive construction/repair ship would follow once the two warships had guaranteed the safety of the JP. Finally, *Dagger* would make transit and the small fleet would continue in concert to the next JP. It was highly unusual to scout the jumppoints in this manner, so far inside Warner territory, but the Commodore was taking no chances. Once the flotilla reached the Antoc system, *Foundation*'s assignment would be to begin the construction of a permanent jump gate at JP1 while the warships began to clear the system's unwanted visitors from their base in the asteroid belt and, more importantly, locate *Pathfinder*.

For this jump, however, *Yargus* would not be using the jump gate. They would be performing the final testing of their new jump engines. This would involve a self-powered jump from a set distance from the plotted jump point. *Yargus* would trail the rest of the convoy except *Dagger*, who would still go in last.

The tension notched upward as the time approached for *Yargus* to make her attempt. Everyone was anxious after the disaster at the previous gate. Her engines had shown fluctuations during the previous jump, which had been made using *Yargus'* own engines. The chief engineer thought she had isolated the problem and they had completely rewired the power coupling to the jump circuit.

As *Yargus* had finally begun its jump from Gregor to Betre, the rewired power regulator had failed and sent the entire charge from the capacitors into the engines. The new engines were not designed to handle that much unregulated voltage and they showed their displeasure with a small explosion. Of the seventeen men in the engine room, only two were close enough to the explosion to have been killed in the blast. Nine others received flash burns of varying severities.

From *Dagger,* the flash had been brilliant enough to be seen clearly from their position near the gate, almost 500 kilometers away.

It had taken the combined crews of the two ships nearly six hours to stabilize *Yargus* enough to jump using the power and guidance of the jumpgate itself. They had used the two days in transit across the Betre System from JP1 to JP2 to make repairs. All that remained was to see if those repairs would hold up.

"One minute to jump," reported Davis from communications as he relayed the message from Ramirez.

Commodore Brighton just nodded. Captain Johnson held her breath as she watched the display for any signs of problems on *Yargus.* She saw none. In fact, she saw no changes at all on any of her displays.

* * * * *

"What the hell is going on down there? Where are my engines?"

Burkhalter listened to Captain Ramirez on the vocom and had no answer for him. Everything checked out. All her boards were green.

She had watched her people move away from the engines as they were activated as if they expected another explosion, but then there had been nothing.

"We're working on it, sir."

She switched off the vocom so the captain would not hear that she had no clue what was wrong.

"Wilson, check all those connections."

Wilson's voice had a squeak as he responded.

"All of them, ma'am? There are over a thousand..."

"I know how many there are. Get your team moving."

"Yes, ma'am."

"Jackson, check the runs on that console."

"Yes, ma'am."

"Truman, make sure there is power to those capacitors. But be careful with those, they might be hot."

"Yes, ma'am."

"Burkhalter!"

Damn! She looked at her chrono. Two and a half minutes. She thought it would have taken the captain at least five minutes to get down here.

"Yes, sir?"

She wiped any evidence of her anger from her face.

"We have new orders from the flag."

Captain Ramirez was making no effort to hide his anger.

"Switch back to Gravitas engines and turn control over to the jump gate control platform. They will transfer the ship into the Antoc system."

She let her shoulders slump. They were being towed into another system like a helpless sack of fertilizer. At least she didn't kill anybody this time.

CHAPTER 10

Dagger - Antoc System
24 August

"Sensor data coming in from *Avram*, ma'am," Petty Officer T. K. Hiramoto called from his seat at the communications console. "Scans are clear. Nothing in the immediate area.

"No signs of *Pathfinder*, ma'am," he added after a slight hesitation.

"All is as we expected. Order *Yargus* to begin pursuit on course 76.0 by 0. Tell them to make their best possible speed per original plan," Brighton directed to the communications officer.

"Aye, aye, sir," Hiramoto responded.

After a few moments, the communications rating turned to Brighton, "*Yargus* responds: Orders understood, following Plan Alpha."

Brighton nodded and moved back to stand in front of his console.

"Orders to *Avram*," he continued. "Proceed to agreed coordinates and begin the search for missile systems and possible enemy base. Beware of possible enemy warships. Consider all ships to be hostile until positively identified. *Pathfinder* is the only other possible WSN ship in the system."

Hiramoto turned away and pulled the exact wording out of the running ship's log to drop into the outgoing comm packet.

There were four very different vessels that had entered the Antoc system under Brighton's command. *Dagger* was the second largest, but easily the most powerful, of the four and, as such, was the flagship of the flotilla. She was a *Rapier*-class light cruiser; designed from the core outward as a fast, heavily-armed combatant to stand toe to toe with larger ships in a missile engagement.

Nothing larger than a light cruiser had been built by any of the Families for more than half a century as the needs of each Family dictated the reality of the shipbuilding agenda. There was a greater need for smaller ships than for the larger, more ponderous, heavy cruisers and battleships. The Families, however, would not give up their missile capacity, so the light cruisers continued to be produced. There had been no large ship to ship missile engagements for nearly a century, but the WSN was following an unfortunate traditional pattern. They were still constructing ships designed to win the previous war.

The heavy cruisers that remained were left over from that first interstellar struggle. As a consequence of that first experience with space warfare, one Family had been destroyed and the allied Combined Fleet had been created, where missile platforms were king. Now all militaries were built to that ideal. The light cruisers were more suited, by size and speed, to defending against pirates and protecting shipping, but still retained the firepower to mix it up with their more powerful cousins. The Treaty of Dallas, which had ended the Vector Rebellion, had also created the Combined Fleet, with the end result that each Family was restricted on the number of warships they could produce. One of the requirements of the treaty was that each Family must contribute one ship of equal size and capability to the Combined Fleet for every warship they commissioned. This was to limit the power of each Family to a level that was lower than the combined might of the Ruling Council. Warner, with the most systems to protect, had the most ships in its own navy, and consequently had contributed the most to CF.

Combined Fleet retained the bulk of its fleet strength near Earth. It was sent on patrols in smaller contingents and occasionally responded to larger events as needed, but it was essentially the hulking specter behind the Council to enforce the peace. Warner retained the majority of her heavy cruisers near their forward base at Gateway System, two jumps away from Earth.

While *Dagger* was a powerful warship, she was still only a light cruiser. As the most powerful component of Commodore Brighton's flotilla, she was the natural choice to send in after the DaGama missile site inside the asteroid belt, with *Avram*, a much lighter corvette, in support.

Foundation was a construction ship, five times more massive and thus more ponderous than *Dagger*. Designed to function independently, far from base, she had only light armament and armor but she was never intended to be part of any fight. Under Plan Alpha, she began moving into position at the jump point. Her job would be the construction of the jump gate and generators that would allow them all to get home.

Avram was a system patrol vessel, usually referred to as a corvette by the generous, and as a "paper airplane" by those less so. Destroyers had been referred to as "tin cans" for longer than navies had been out in vacuum. This was because they were built with an eye to acting as screening elements for larger warships; simple and ponderous. Corvettes, by contrast, were built to be faster and more agile than anything else in an engagement. The tradeoff for maneuverability is mass, which meant a corvette carried only minimal armor and armaments; greatly skewed toward light-speed lasers rather than missiles, since missile bunkers would add mass.

Like all of her class, *Avram* was designed to protect shipping from point to point within a star system, but rarely moved from that assigned system. They were dangerous opponents within their sphere of engagement, but not designed to take on anything heavier than themselves. Several had disappeared over the years without any explanation, which indicated that there were larger vessels out there in the hands of those unfriendly to the Ruling Council.

Yargus was the final component of the small flotilla. She was a destroyer of the *Risea* class, sixty-two years old and small compared to modern destroyers, but she was freshly out of a refit that made her irreplaceable. She contained the only existing jump engines outside of the *Pathfinder* and *Vanguard* prototypes. *Pathfinder* had been based on the *Risea* hull as well, and the two ships comprised two of only three remaining hulls of that type still in service. *Yargus* had not been as extensively reconfigured as *Pathfinder* had been, and she retained most of her offensive armament, losing only one of her aft missile tubes and a rear laser turret to the jump engine capacitors.

"*Yargus* acknowledges orders, ma'am," Hiramoto reported. "She is heading out on *Pathfinder's* last known course."

"*Dagger* and *Avram* will execute search plan delta. *Avram* will move directly to the asteroid belt and begin a search for missile platforms and tracking systems. *Dagger* will head for the planet Le Vesconte and enter orbit," Brighton continued.

Johnson heard a sharp intake of breath from the engineering console behind her where Derrick Mackey was seated, but she did not look. She knew the source would be the large NCO who would be having the same emotional reaction to that name as she was having herself.

"Is that A3, sir?" asked Johnson with a surprised tenseness in her expression.

"Yes, Captain. Before leaving Earth, I petitioned the Board for the right to name the planets of this system. That permission was granted. I believe you will find that your database has the correct names attached."

She pulled up the chart of the Antoc system and it was just as he had said. In place of the numerology that had been a part of their existence while trying to escape from the system, she found that all the planets now had real, honest to goodness, names. A3 was now labeled Le Vesconte.

Commodore Brighton turned to his former ship's bosun and said, "I'm sorry, Mr. Mackey, but I could not let your name stand on La Paz. Fujinami felt more appropriate under the circumstances."

"Wha... yes, sir. Very appropriate, sir." The dour noncom said while trying to hide his emotions. Drew Le Vesconte and Jens Fujinami had been crewmembers on *Pathfinder* and had accompanied Commodore Brighton into exile on board the ship's launch, *Vanguard*. They were the only members of the crew who had not lived to return to Earth.

"Captain, I believe you were going to order a course change…" Brighton gently reminded his flag captain as she stood rigidly working through her memories.

"Yes, sir," she said, not allowing her face to register her feelings. "Helm, make your course 104.4, same plane, at forty gravities."

"Yes, ma'am"

"Comm, one final message to *Yargus,* if you would?" Brighton said, "Proceed at best possible speed. Expect at least six weeks to intercept. *Dagger* will join at first opportunity. Expect rendezvous before overtaking *Pathfinder.*"

"*Yargus* signals, 'Understood.'" Hiramoto replied after a few moments.

"Captain, please signal Major Chowdhury to prepare a squad for planetary security. I will be revisiting our final landing site for a short period of time and would appreciate her company."

"Yes, sir," she responded in a surprised tone. She had not known of this side trip to the planet. She contented herself with the knowledge that everything the commodore did had a direct purpose. "Shuttle one is prepped and ready to go."

Several hours later, *Dagger* entered orbit around Le Vesconte and began scans to determine the proper site. Finally, it was isolated and the landing team was assembled.

WSN assault shuttles could hold twenty-four fully equipped Marines when configured for planetary insertion. Apparently, twelve was the number that Major Chowdhury had decided was appropriate to protect the commodore during this mission, because that was the number that stood at attention in full armor as he and Captain Johnson entered the boat bay. Johnson was also startled to see the faces of every member of the *Vanguard* crew currently aboard *Dagger.* Lieutenant Leonard Ward, Tim O'Neill and Elle Williams had not been medically released in time to ship out with *Dagger,* and Kieran Delacoeur had decided to take his accrued leave. With those exceptions, all of *Vanguard*'s former officers and crew were assembled in the boatbay.

"Commodore on deck," came the call and everyone snapped to attention in unison.

"At ease," Brighton responded immediately, before surveying the assembled group with a wry grin. "It would seem that word of our excursion has gotten out. We have some unfinished business to attend to."

Turning to Major Chowdhury he added. "Drop the second shuttle with a squad of Marines at our first landing site. Their orders are to clear the planet of any hostile troops they locate. Take prisoners if they can safely do so, but their safety is of the highest priority."

"Yes, sir," she replied and began making the calls that would prep the shuttle and the Marines.

"Captain," he began as Johnson turned her attention back to him. "Your place is here with your ship, so I won't invite you to join us, but I felt that you should be here with us now." Turning to address the remainder of the crew, he continued. "We are beginning the first step of our journey back to *Pathfinder*. We were forced to make hard decisions and compromises to enable our safe return to Earth. One of the least significant, from a physical standpoint, yet also one of the greatest, from an emotional standpoint, took place on this planet. Of the greatest, I am of course referring to Drew Le Vesconte, who sacrificed his life here. As for the second, if you will all recall, I asked each of you to abandon any nonessential items in order to lighten *Vanguard* to a point where escape was feasible. Everyone did so without complaint, though many of the items had deep personal meaning or sentimental value. I also made you a promise that we would retrieve them at the earliest opportunity. We are now going to redeem my promise. Any who wish to accompany me are welcome to do so."

The entire group loaded quickly onto the shuttle and Captain Johnson watched in silence as the hatch was closed and sealed. Her heart ached to be with her friends as she moved to the edge of the cavernous boat bay and watched the shuttle drop out of its cradle and plunge toward the surface, while the second shuttle loaded with Marines. She watched through the transparent shield until the dot was no longer visible, then made her way silently to her quarters. She just wanted a few moments alone for reasons that she could not define.

CHAPTER 11
Le Vesconte
25 August

The stillness of the newly renamed planet Le Vesconte was shattered by thunder from a pair of assault shuttles as they tore through the atmosphere toward their destination. The first shuttle hovered over a large clearing while the second settled to the surface and disgorged its Marines.

The soldiers scattered in various directions with what appeared to be random chaos but was in fact a well-designed and executed maneuver. They secured the entire perimeter of the clearing before the first craft could begin its own descent. The matte black shuttle lowered ponderously to the ground in the center of the clearing next to the smaller Marine insertion shuttle and the large rear assault ramp dropped to the ground. Commodore William Brighton exited, followed by Major Chowdhury and the remainder of the officers and crew. Those who disembarked represented only a fraction of the crew from *Dagger*, but they were nearly all of the remaining crew of the exploration launch named *Vanguard*.

Ensign Jherri Roberts looked at the tree line in front of them and then hung her head without expressing the thoughts that clouded her features.

"You did a good job here," she heard Major Chowdhury say from beside her. "You have nothing to feel ashamed about. Most junior officers would be lucky to perform half as well as you did."

"Thank you, ma'am," she replied automatically, but the Marine was already moving off to the trailhead. Roberts shook her head at the thought of receiving praise from the taciturn security officer. She had rarely heard her say anything positive about any of the ensigns during their sojourn on *Vanguard*.

The group moved off down the trail following Chowdhury and Brighton. No one spoke, as each was caught in their own private thoughts and recollections. This planet meant different things to each of the members of the group.

Le Vesconte, once known as Antoc-A3, was just as Roberts remembered it. The dark trees covering the trail like a tunnel and obscuring any view more than a few feet from it. She walked slowly, just behind the other two ensigns, Mitchell and Hayes. Their nonstop banter was a comforting reminder of the closeness they had all shared during their escape from this system, and also, all of the tragedies they had endured. These two could be exasperating in a way that only family could be. She knew their thoughts and what they would say before it came out of their mouths. She laughed suddenly and they turned to face her.

"What?" Hayes asked

"Nothing," she replied. "Just a stray thought."

"Oh, no," Mitchell chimed in. "You're not getting off that easily. Come on, out with it."

"I was just picturing you two strapped to the ceiling after that slingshot burn around Antoc B. Like two flies caught in a spider web of their own devising."

Color crept up their necks and onto their faces like a rushing tide. They turned and began to walk again without any further comment.

"It can't be any worse than you bursting from the forest with rifle blazing, running for all you were worth and screaming your name at the top of your lungs. You looked like a total fruitcake," Mitchell said finally, with a chuckle. "Then Chowdhury came charging out behind you and I thought she was finally declaring war on ensigns and you were trying to keep from becoming her first victim."

Roberts went silent again as his comment brought back the events of that day, which her levity had only just managed to push away. That had been the worst day of her life and even though it had happened half a planet away from this spot, the vegetation was too similar to ignore. The feelings she had been pushing down since that day were trying to bubble up and overtake her. She had been sure that she was going to die. She had been trapped in the center of a clearing with only the cover of a small hole to hide in and several people shooting at her. Drew Le Vesconte, the quartermaster, was already dead, a hole blown through his chest. She had pulled him into the hole with her before she had realized he was beyond her help, his sightless eyes completely different from the vibrant blue gaze he had turned on her just five minutes before. It had been completely unexpected. *One minute we were sharing a meal with the scientists and the next they were trying to kill us.* Then one of the ersatz scientists had jumped up and sprinted the few meters that separated them, trying to take her prisoner. She didn't know what his plan had been. Hostage? Prisoner? Play-toy for later? All of those thoughts had flashed through her mind as he approached. She had had nothing but a few large rocks with which to defend herself but she had been so angry that she had grabbed one and prepared to do as much damage as possible. Then his blood and brains had sprayed out across her face and he had collapsed in a boneless heap on top of her.

"Hey, are you all right?" Hayes asked as she fell slightly behind them in trooping through the forest.

"Yeah," she replied. "Just remembering the last time we were here."

"What did happen at the feast?" he asked, suddenly looking concerned. "You never did say."

"You saw what happened at the ship. It was pretty much the same at the party."

"They attacked the ship about an hour after you guys left," Hayes said softly, his eyes down.

Whatever he had been about to add to that statement was cut off as the group entered a clearing and a shout was raised by those in the lead.

"Here it is, sir," called Mackey.

"Get it lowered down, if you would, Master Chief," the commodore replied.

"Aye-aye, sir."

The stocky NCO quickly organized a team and got all of the containers lowered from the tree branches down to the pale yellow grass.

Everyone stood silently looking at the sealed containers as if unsure what to do next. Kara George broke the tableau by racing to the second container, releasing the bindings and grabbing a small porcelain ballerina. She looked at it, clutched in her hands, and quietly began to weep.

"I'm sorry, sir," she said, looking up at the Commodore through her tears. "It was given to me by my grandmother. It's… er… it's very special to me, sir."

Everyone started moving forward, as if her movement had opened the floodgates. There was no pushing or shoving, but everyone moved determinedly toward their own treasures.

"Sir," Crewman Alcaraz said as he leaned close to the Commodore, "thank you for getting us back here. This ain't much," he motioned with a large antique snow-globe in his hand, "but my mother gave it to me when I left home at eighteen. I've kept it ever since. She's gone now. It's all I've got left of home."

He shuffled his feet and looked slightly embarrassed at his admission.

"Anyways, thank you, sir."

"No thanks are necessary, Crewman. I gave my word," replied Brighton, looking equally uncomfortable.

"Commodore," Chowdhury called as she walked up the slight rise from the milling group to where Brighton and Alcaraz were standing. Alcaraz backed away with a mumbled "Excuse me."

Brighton nodded to the crewman and turned his attention to the Marine.

"I would like your permission to move to the attack site. We need to ascertain if the DaGamans are still on-planet here and neutralize them if they are," Chowdhury requested.

"Yes, if we are ready to move on to that phase of the landing then please have the crew embark onto the shuttle. As we discussed previously, we will land our shuttle at *Vanguard*'s third landing site while you and the other Marines will land at the terraforming camp. When all is secure, you will call us forward."

"Aye-aye, sir." She snapped a salute and trotted off toward her second-in-command. The Marines instantly started moving and shrinking the perimeter. The crew started gathering their belongings and moving in the direction of the shuttle. Brighton moved over to the containers and transferred the remaining items into the largest. He then sealed it and hefted it to his shoulder to transport it back to the ship. It suddenly lifted off of his shoulder and he nearly overbalanced at the loss of the load. Mackey strode past him with the container tucked negligently under his arm without saying a word.

They spent nearly two hours moving to the chosen landing sites on the far side of the planet. Moving eastward, they went from mid-morning to late afternoon, local time. The crew had barely landed and disembarked when they received the 'all clear' to move forward. As they moved across the clearing from the shuttle to the trail there were still numerous visible signs of the short, violent conflict that had raged here only a few short weeks before. Brighton led them down the well-marked trail toward the site of the ill-fated feast. Roberts remained silent as the memories flooded back again and she felt the fear grip her that had nearly consumed her during that battle. Hayes and Mitchell walked behind her. Even they felt the mood and remained uncharacteristically silent.

As they entered the clearing, Chowdhury stepped forward and saluted.

"The area is secure, sir."

"Very well. What did we find out?"

"There is no evidence of either the original terraforming team or the DaGaman occupants. They sanitized the area pretty well when they pulled out, but it looks like they left in a hurry. There is no evidence of any remains. They must have taken our dead with their dead and wounded."

"All right, if there is nothing further to--"

"Begging your pardon, sir," Mackey interrupted in a quiet voice. "Would it be possible to say a few words for Mr. Le Vesconte?"

Brighton looked over to the massive noncom and the group slowly gathering behind him. He looked over the faces that had become so familiar to him during the trials of the last few months: Kara George, Ricardo Smith, Steve Long, Mitchell, Hayes, Roberts, Alcaraz, and the rest of his surviving crew. They each looked up at him with expressions very similar to the ones they had used at the beginning of their trek. They wanted him to make sense of everything. They wanted him to let them know that their sacrifices had been worth it.

"Very well, Master Chief, please gather the crew." He noticed out of the corner of his eye that Chowdhury motioned her Marines to fan out and guard the clearing.

"We never had the chance to properly mourn for those friends we lost along the way," he started simply. "They deserved a better fate than that which they received."

There were some nods as the group pulled slightly closer together to be able to hear his quiet words more clearly.

"I don't know what plan is laid out for each of us by our Creator and I don't know much about the faiths or beliefs of Drew Le Vesconte or Jens Fujinami, but I know that they both willingly laid down their lives for the rest of us. As I think back on tiny Jens as he attacked an armored opponent with only a sharpened stick, I feel that it is possible I never knew him at all. He was always laughing and seemed just slightly out of touch with the world the rest of us live in, but he was ferocious in his defense of our lives. Drew knew the risks we took coming here into the camp of the enemy, but he played his part faultlessly. He never hesitated nor drew back from the possible conflict."

Brighton stood silently for a few moments as all thoughts were turned inward, each person reliving their own memories of those fateful days.

"I am proud to have had comrades like these. If any of you have feelings or thoughts that you would like to share, please feel free to do so."

The silence was profound for several moments before Clémence Quèneau spoke. "Jens treated me like a daughter and he was always trying to show me things. I remember once he came into the galley with a four-centimeter bug that he had caught in the interstitial spaces during his off watch hours. He was smudged and dirty but he sat there talking to the bug as if it were a long lost friend. His world was a world of constant wonder and amazement."

She stepped back a few paces to indicate that she was finished speaking. The silence descended again but only for a few moments before Master Chief Mackey began speaking.

"I remember once, on leave at Idyll, Le Vesconte and I had gone out to dinner," the bosun began with a smile on his face. "We were assigned to *Katanu* on system patrol and we had just returned to the station after a deployment of five months. These two bruisers off a freighter came in and started to cause problems. They had been drinking and thought the station personnel owed them a free meal. Things kept getting louder and uglier until Drew stood and walked over to their table. They were both easily twice his size. I couldn't hear what he said, but pretty soon they were laughing and ended up arm-wrestling for their dinner. They were so drunk they couldn't even stay on their chairs. Drew beat them both, then walked out arm-in-arm with them, laughing. That was the way he was. He always looked for the humor in every situation and everyone was his friend." The large NCO backed a single step to show that he was done. There were more smiles on faces now than tears. Everyone was lost in the reverie of their own memories.

"I have never told anyone this before," Roberts began in a voice barely above a whisper, "but without Mr. Le Vesconte, I would be dead. I was sitting next to him at the feast," she said, never raising her tear-filled eyes from the ground. "He pushed me out of the way. I landed in a hole and by the time I pulled him in, he was dead."

Everyone stood looking expectantly to see if she would continue, but she stood unmoving. The light mood which had taken over was again darkened. There were no smiles of happy memories.

Commodore Brighton looked over the assembled officers and crew. None looked willing to break the new mood of the gathering.

"Is there anyone else who would like to speak?" he asked.

When no one responded to his question, he turned to his security head, who was standing at the edge of the group with her helmet tucked under her arm.

"Major Chowdhury, radio the pilot of the other shuttle and have them retrieve us here."

"Yes, sir."

"Are you convinced that this planet is secure?"

"Yes, sir."

"Are you ready to proceed against the asteroid base?"

"Absolutely, sir," she said with a look that could only be called feral.

CHAPTER 12

Dagger
25 August

The cavernous boat bay of WNS *Dagger* was a flurry of frenetic activity. Equipment packs, storage bins, tool kits, and various odds and ends were cluttered about the deck plates as a result of several ongoing jobs. Loud voices shouted instructions and passed information from one end to the other, sometimes requiring a relay to get the message to its recipient.

Chief Warrant Officer 2 Stephen A. Long paused a moment in his labors to rest and catch his breath. While he did so, he considered the differences between this mammoth hangar and the last boat bay he had been in.

The last time he had been in such a space was not a memory he recalled with any fondness. It had been the last sight he'd had of *Pathfinder* before being evicted from that ship and it had been the beginning of his nightmare voyage on *Vanguard*. The very fact that he and Derrick Mackey, at the other end of the crate they were relocating, had to stop and rest was a testament to how hellish and punishing that journey had been. Long's trousers' waist measured less than 80 cm for the first time since he was fifteen, and having the energy to work long hours, even long minutes, without a break was a vague memory.

The major difference *Dagger*'s bay boasted, apart from the sheer size, was that the boats it serviced were stored inside. *Pathfinder* had, in its former incarnation as the destroyer WNS *Risea*, mounted two assault shuttles to its ventral surface, which were accessed via an enclosed ladder and companionway from the centrally located boat bay. By the time she was reconfigured as a survey/exploration ship and renamed, the larger survey launch, *Vanguard*, had occupied both of the two slips under *Pathfinder*'s belly.

Dagger had never been anything other than the *Rapier*-class Light Cruiser she currently was. In fact, being only five years from her commissioning, she was a dream come true for a maintenance chief like Long. *Rapiers* were meant to serve a number of roles in the Warner Naval Service, and the four assault shuttles, two launches, and single armed pinnace, each visible in their proper berth from where Long had stopped, afforded her the flexibility to fill each of them. One of those roles, attacking an enemy stronghold, was the cause of the current activity surrounding them, directed by an ensign of average build, average height, and unremarkable features.

As the officer in charge of boat bay operations, Ensign Polunu Keone, intended to stay on top of any changes or assignments in his command area. Commodore Brighton had returned from whatever business he'd had on the planet, and *Dagger* was even now speeding across the system to rendezvous with *Avram* in order to eliminate any threat from a hostile military base in Warner space.

This meant it was Keone's job to outfit two assault shuttles for boarding action, two for mine-laying, and the pinnace for space combat. That should cover the requirements for whatever situation *Avram* reported. Be prepared for anything, was Keone's motto. It was difficult to manage, though, and attempting to do so was not a recipe for brightening Keone's mood. Not that his mood was ever all that bright.

"You waiting for a personal invitation, Long?" the young officer asked with unmasked disdain. "Or is it that since you're a media celebrity now, you brought along a personal retinue to do your work for you?"

Keone had not had Long working for him for long enough to get to know him, so it was understandable that he should approach the situation in exactly the wrong way to elicit the response he wanted. Were he more experienced with Long's personality, he would have known that the proper way to deal with the maintenance Warrant Officer would be to tell him what needed doing, and then to stay out of his way. Frequent reminders or constant supervision, even during the best of conditions, would cause Long to dig in his heels more firmly than any mule ever to see the light of day. This, however, was far from the best of conditions. As a warrant officer, Long normally commanded crews and organized technical repairs and maintenance. Rarely was he assigned grunt work. He neither enjoyed the experience nor reacted well to it. The fact that he and Bosun Mackey had both been assigned this same menial task showed that some malice was at work at some level.

"Stick it in your ear," Long suggested helpfully. After a long pause he added, "Sir."

"Are you being insubordinate, Long?" Keone demanded hotly.

Mackey jumped in before Long could make the situation worse. The last thing Long needed was to face a Captain's Mast for insubordination charges. Again. "Don't worry, Ensign. We just needed a little breather. We'll have things loaded on schedule." Mackey's grin and placating tone mollified the ensign somewhat.

Not so for those who had begun gathering. "I wouldn't let them slide if I were you, Ensign," Master Chief Durrant said from the forefront of a cluster of *Dagger*'s crew, walking over to join their community against outsiders. "Brighton might have allowed a loose ship, but it's time they learned there's no slackness on this one."

"Well, now that the Great Savant has shown how much he knows of *Commodore* Brighton's personality and command style," Long gave exaggerated stress to the rank, "we should now humbly bow before jumping to do his bidding! And we'd best be quick about it," he added in a growl. "I hear we also need to take on a load of ice exported from Hell."

"Easy, Long," Mackey said. "They're just expressing their opinions. Nothing wrong with that."

"And my opinion is that you should quit stalling. Then we can blow *Pathfinder* out of the void and get back to Earth," Keone said. Part of Keone's foul mood, and the immediate dislike of the interlopers generally, was that the entire crew of *Dagger* had been due for six weeks of liberty when they returned home from a twelve-month deployment. Instead, they were the handiest ship to assign for immediate return to duty when Brighton arrived and set the Naval Board scurrying for a pursuit force. In their eyes, the new additions to their crew were to blame for the fact that they were back inside this metal shell instead of enjoying some down time.

Keone also had personal reasons for disliking the changes; specifically the change in command. He had nothing personal against Brighton, it was his choice in flag captains that so irritated him. Keone had graduated from the Warner Naval Academy four years earlier, and since that time he had remained firmly at his same rank. He never seemed to be the right man for the job at any of the postings he had applied for.

Captain Johnson was only five years out of the academy, he remembered seeing her here and there, and yet she was now three grades his senior. His career had stalled right out of the gate, but every break in the universe had seemed to fall into Johnson's lap. The injustice of the situation burned him inside.

Injustice nettled Mackey as well, but it was the unfairness inherent in Ensign Keone's words, and he was quick to respond, "Not everyone that stayed on *Pathfinder* is guilty, Mr. Keone. Several innocents were left through no fault of their own, and many of those on board are or were my friends. There is no need to look for heartless retribution here, only justice for the guilty."

"Yeah, kid, if I want you or Durrant to express an opinion, I'll beat it out of you," Long said. Mackey looked at him carefully. It seemed like Long was spoiling for a fight here, but by Mackey's count, it would be eight against two. Four, he amended, as Roberto Alcaraz and Claire Paul unobtrusively joined their side of the gathering. Still not good odds if things got ugly.

Durrant shrugged, ignoring Long's threat. "Well, it's too bad for them, but they're all going to wind up dead. Whether they deserved it, or they're just collateral damage won't matter in the end. Get over it, and let's get on with it."

"Now that you and the *commodore* are back from your sightseeing tour of the planet, those shuttles need to be outfitted all over again. Now hop to it." Keone sneered as he mentioned Brighton's rank, causing Long's hackles to rise. On an average day two months before, Long would have been happy to join in the general derision of authority figures, but a lot had changed for him in that intervening time.

It had been brought to his attention, rather forcefully in one case, that he owed his life to the leadership and abilities of one William J. Brighton, and that realization, once it had sunk in, had a profound effect on his outlook in a number of areas. Everyone who had endured the torturous crossing of the Antoc System aboard *Vanguard* was much nearer family now than simply shipmates, and nobody got away with slighting Long's family.

"You know, Keone," Long began, deliberately not mentioning the man's rank, "when I earned my warrant, there was a requirement to demonstrate both ability and intelligence. Had I known there was no such constraint for a commission, I would have taken the other track and saved all that effort."

Keone's face turned red, but he made no response. Bill Sydow picked up the challenge for him, though.

"Pointless to argue with him, Ensign," the crewman said, as a few more gathered around them, sensing the tense mood. "After all, Brighton's got a commission, and anybody with either of those qualities would have been able to manage to keep his ship from being stolen." A grumble of agreement and more than a few chuckles could be heard as others dropped what they were doing and gathered to add their thoughts on the issue.

"Yeah," an unidentifiable voice in the back added, "just as pointless as stopping for a look-see at an uninhabited planet when there's a base full of squatters in the system." This comment touched off another brief laugh, which Keone joined.

"Honor is what took Commodore Brighton to that planet," Mackey's deep base cut the other's mirth short. He raised himself to his full height, fifteen centimeters above anyone else in the bay, as he spoke. "It was a personal duty, and I know none of you were aware of his reasons, so I'll forgive your comments. The commodore gave his word to retrieve items left there at the first opportunity, and that is exactly what he did."

"So, lost his ship then he lost his gear, too? Sounds like a *fine* officer to me. A real shining star, among the finest in the Warner Navy!" Crewwoman Kjira Soderburg's voice dripped with sarcasm.

"And what was down there worth the trouble to go get?" Keone asked now.

"Wouldn't have mattered if it was something as worthless as your mother's virtue. He gave his word," Long fired back. The verbal missile found its mark and again Keone reddened, now advancing on Long in response.

Mackey stepped between them quickly. "Hold on now," he said in a longsuffering voice. "That was uncalled for, Steve. You should apologize to the ensign. And you, Ensign, should never question the honor of a superior officer." The rebuke was out of line, from any non-com to an officer, even from the Bosun. Mackey knew that, but he also knew Steve Long well enough to know that he would not back down first. To stop this from escalating, both sides were going to have to back away from the conflict.

Neither side seemed particularly interested in backing away.

"Face it, Long," Durrant said mockingly. "The man had his ship taken away from him. He's got to be twelve different kinds of idiot to let that happen." Unashamed laughter filled the bay.

Unlike Keone, Long's face did not redden, but his eyes narrowed and a subtle change in his footing prepared him to either lunge forward or absorb the force of someone else's attack. Mackey noted this and knew that Steve was not likely to let someone else land the first shot. Alcaraz was tense and had angled his stance to cover their right flank, clearly expecting that this was going to come to blows, and quickly. Claire had positioned herself to the left, and he noted that she still held the heavy wrench she had been using to attach an umbilical to the pinnace.

The hecklers seemed oblivious to the preparations of Long and the others, or perhaps they didn't care. They had the small group outnumbered nineteen to four, so what could the former *Vanguard* crewmen do but stand there and take it?

Chief Giovanni added fuel to the well-stoked fire. "*Pathfinder* is probably in better hands now, anyway," she grinned. "The people in charge now were at least smart enough to take the ship away from Brighton. Although, I suppose it wouldn't be too hard, considering his limited mental ability." This brought forth more gales of laughter.

"You don't know what you're talking about," Claire forced out through her clenched jaw. "If a better officer ever lived, I never heard of him."

"You need to read more, Paul," a derisive voice in the back pronounced. Laughter echoed the suggestion.

"Brighton is nothing but a swaggering, overbearing, tin-plated dictator with delusions of adequacy," Keone said. The ensign smiled as this brought more laughter.

Uncaring of the consequences, Long took a big step forward, until Keone was within arm's length of him. An anxious look crossed the ensign's face and the enemy formation tightened up, expecting they had finally provoked them into a response. Instead of attacking, though, Long simply looked down at the officer.

"Boy," he practically spat, "would you like to…rephrase that, or would you prefer to swallow your teeth?"

Again, Keone's lack of experience dealing with Long caused him to say exactly the wrong thing. "You wouldn't *dare* strike a superior officer!" To emphasize his belief he jutted his chin out defiantly.

Mackey recognized his blunder before the words were out of the unfortunate officer's mouth. He knew what Long's reaction was going to be, and also knew what the consequences were likely to be for his friend, given his record. Mackey could think of only one thing he could do to save Long.

A quick step to the left blocked Long from advancing, and Mackey's big right hand came flashing out to smash into the protruding jaw. Two teeth flew free and made a small sound as they hit the decking just before the thud of the ensign's now inert frame.

CHAPTER 13

Dagger

25 August

Silence dominated the large briefing room, broken only by the slight swish of fabric with each step as Lieutenant Commander Johnson, Captain Johnson while on her own ship, paced in front of the assembled men and women. There was not even a click to accompany her steps; the day-to-day undress uniform used soft-soled deck shoes rather than the hard-soled boots of the dress uniform.

This situation was not new to Steve Long. Generally speaking, you could measure the anger of the commanding officer by how long the silence lasted. A longer period of silence indicated more anger that the said officer wanted under control before speaking. Johnson had just passed six minutes of pacing, which was nearing a record period, in his experience.

Johnson turned then and finally looked at the assembled group. She still did not speak, only stared at each person in turn. They were a mess.

The two groups that had been fighting had divided themselves by instinct upon entering the room. The former Vanguardians had entered first and were farthest from the door to sternward. Everyone had several bruises just beginning to become visible. The left side of Ensign Keone's face was swollen. Three wore arms in slings, and two of those had casts to immobilize broken bones, the third keeping the shoulder immobile instead. Bill Sydow, she knew, was wearing a chest brace to support very tender broken ribs. Derrick Mackey wore a uniform tunic that was nothing more than tatters, but he had arranged it as modestly as he possibly could. Despite their obvious injuries, each of them stood rigidly at attention, with their eyes fixed across the conference table and on the far wall.

As Long's count reached seven and a half minutes, Johnson finally broke the weighty silence. "There is no excuse for fighting on board a Warner ship.

I want to know who started it." Her voice was sharp and firm but aside from two people who flinched at the sudden sound, there was no response.

"I'm waiting," she said, but still received only blank, impersonal stares. She stopped in front of one and turned to face her.

"Vreeland, who started the fight?"

The crewwoman addressed answered quickly, "I don't know, ma'am."

Johnson continued to stare for a moment, trying to look menacing in order to elicit the information. She felt a little ridiculous, knowing that the woman was both older and bigger than she herself was, but she kept any indication of her personal inadequacy from reaching her face. She made a mental note to see if Major Chowdhury would give her lessons in intimidation.

She turned and paced back the other way until she came to her number one suspect. "All right, Long. I know you. You started it, didn't you?"

"No, Captain, I didn't," he said. His eyes were steady and there was no hesitation in his words. Still, she was sure he'd had plenty of practice at lying convincingly.

"Well then, who did?" she followed up immediately.

"I don't know, ma'am." A slight hesitation this time.

"I don't know, ma'am," she mimicked slowly, not believing him for an instant. "I want to know who threw the first punch."

Long made no move to provide the information, and her gaze swept up and down the line, looking for someone who would give her a straight answer. She found her target one man to her right.

"Mr. Mackey, you're an honest fellow."

"Thank you, Captain."

"You're welcome, Bosun. Now tell me who threw the first punch."

There was no hesitation before he confessed, "I did, Captain. I threw the first punch."

That caught her completely flatfooted. The man had to be lying; he wouldn't hurt a fly. Yet she had just finished praising him for his honesty. How could she turn around and accuse him of falsehood and press him for the truth. He must feel it necessary to protect whoever did start it.

She turned instead back to Long, still her prime suspect. "Is that the way it happened, Mr. Long?"

"The very idea is laughable, Captain. I cannot think of any set of circumstances which would lead Mr. Mackey to start a fight."

"Are you telling me the bosun is lying?"

"No, ma'am. Mr. Mackey does not lie."

"Then how do you resolve your two statements, which contradict each other?"

"The bosun must have made a mistake, ma'am."

"The bosun must have made a mistake," she echoed slowly. This was getting her nowhere, and it was making her look ineffective, downright silly even.

"All right. You're all confined to quarters pending an investigation. Dismissed." The crewmen filed out smartly, backs stiff and steps uniform. "Ensign Keone, you will remain."

Polunu looked resigned to whatever fate awaited him as he stepped out of line and let the others file out of the room. Although, she had to admit, it was difficult to tell what sort of expression he wore, with the distortion of the man's features induced by the lopsided puffiness.

When the crew had gone, leaving only herself, Keone, and her executive officer, Lieutenant Grant, she sat down at one of the chairs and motioned for the ensign to take the one across the table. Grant slid into the one adjacent to her without being directed. He seemed distinctly unhappy with the events of the morning, as well he should. It was his duty to manage the personnel of the ship, and to take care of any friction without bringing things to the attention of the captain. A general melee involving everyone on duty in the boat bay could not be ignored by the ship's commander, however.

"All right, Ensign Keone, why don't you tell me how the fight started?" Captain Johnson prompted calmly.

"I didn't see any of the fight, ma'am. I was unconscious through the whole thing." His words were a little difficult to make out. Hopefully, the swelling would go down soon.

"So you were the first person to be hit?" she asked pointedly, leaning forward a bit.

"Yes, ma'am."

"Who hit you?" she asked in a hammered iron voice.

The ensign shifted his weight in his seat, and looked from one superior to the other. Both wore equally grim expressions. "Is this off the record?" he managed to ask.

"No. This is not off the record," she said gravely. "Now answer the question. Who hit you?"

Keone was torn, and that surprised him. By all accounts, he should just confirm Mackey's guilt and let the chips fall where they may. He should be angry enough, and want to get even enough, to be happy about the idea. He certainly was in enough pain that he ought to want revenge. Oddly, he didn't.

Had it been Long who had punched his lights out, he was certain he would not have hesitated. But Mackey was a different story. Everyone looked up to the bosun, literally as well as figuratively, and Long was right about the man's patience. It was hard to imagine him getting mad enough to hit someone. Certainly, he would have to be provoked a great deal to do so. And that was what held Keone's tongue.

He recognized his own guilt at providing exactly that kind of goading.

"I don't know, ma'am. I never saw it coming." The lie came easily to his lips, though he knew it would not fool the captain or Mr. Grant; not after asking if he was speaking on the record.

Johnson's eyes bored into him, but he sat motionless and endured it as best he could. He didn't know why he had ever thought her too young for a ship's command before. Sitting here before "the old lady," it was clear to him that she carried her authority easily and naturally.

"I see," she finally said. "In that case, Ensign, let me tell you whose fault the skirmish in the boat bay was." She stared intently at him until his eyes raised to look at her, and she was sure she had his complete attention. "It was my fault, Keone." He made to disagree, but she held up a hand to forestall him.

"I'm the captain of this ship. I am responsible for everything that happens onboard *Dagger*. Those crewmen were injured because I didn't do my job." Her finger stabbed toward the hatch, then redirected toward her chest.

"At the same time, it was also Lieutenant Grant's fault. He is responsible for dealing with crew issues." She paused to take a deep breath, then let it out before continuing. "And it was your fault, as well."

He flinched at the heat in her words as if a lash had fallen across his back.

"You were there, and you were in command. It was your duty and your responsibility to see that it never happened," she continued, though with somewhat less heat. "I do not know the details of the event, and part of me hopes I never learn them, because if I do, my duty will require me to end someone's career for striking an officer. But you know what happened, at least the events that caused a fight to break out. I'm sure if you play it back in your mind, you'll see that there were opportunities for you to stop the situation from progressing down that road."

Keone stared down at the conference table morosely, no longer able to look his captain in the eyes. "Yes, ma'am," he agreed unstintingly. "I can see that."

He gathered together his courage enough to look her in the eye. "Whatever punishment you think is fitting, I am willing to accept. I know I failed to do my duty."

She nodded once, judging that he was indeed willing to accept responsibility for his own actions. "An appropriate punishment will be given later. For now, you are dismissed to medbay. Doctor Trevino is expecting you for oral surgery."

The ensign rose and saluted. She returned it quickly and he departed.

The weight of her accountability hit her full force then, and she turned to look at her XO. "Do you have any recommendations, Mr. Grant?"

"Jim, please," he entreated.

"All right, Jim. Do you have any recommendations?"

"Not at this time, Captain. Let me think about it, though, and I may be able to come up with something."

She glanced at her chrono. "Very well. You're overdue for your bridge watch. I'll let you get back to that."

He rose and saluted, exactly as Keone had a moment before, and she returned the salute as dismissal. Alone in the conference room, she sat and thought. She really didn't know what the best thing to do was. She had lots of ideas, punishments she had seen used before, but somehow none of them seemed quite right for this situation.

She knew the goal she was after: reduced friction, efficient work, a happy crew. *And bored doctors*, she added. How to reach those goals, though? That was the question. She wished suddenly that she could be more like Commodore Brighton.

He always knew the answers.

CHAPTER 14
DaGama Base
25 August

Major Sheli Chowdhury, WSMC, pulled off the helmet of her light armor and thought that this, quite possibly, had been the easiest operation she had ever conducted. That put her on edge. She didn't trust easy. Ever.

She commanded what were essentially two platoons, designated a dragoon.

She was currently supervising the mop up of the base take-over. She, and her dragoon, were tasked with taking the asteroid base that had been identified during their escape from this system while aboard *Vanguard*.

The take-over had been exceptionally easy. Chowdhury had a basic, fundamental, almost cellular level distrust of anything easy. Easy is unnatural.

Base personnel were being escorted to the make-shift POW 'camp' that had been set up in the main hangar bay. Most of those prisoners were eager to have the corpsmen look them over and get themselves some sustenance. Mostly, they just wanted the food.

When she and her Marines had approached the base installation, two and a half hours earlier, they had picked up a distress broadcast on a repeating loop. The signal was very weak, so it had not propagated much farther into the system. After listening to the voice's repeated request for immediate assistance, she had broadcast a message back. She had given them two choices, immediate unconditional surrender, or a full Marine incursion.

The surrender had been less than forty seconds in arriving and it was followed by a second transmission, not five seconds later, asking how soon it would be possible to get food shipped in once they were prisoners.

When the assault shuttles landed, they met no resistance. Neither did they find all of the base personnel. Many had not responded to the summons to assemble. Possibly they did not even realize the base had been occupied.

Notwithstanding the disorganization, the vast majority of the base personnel had been where she had ordered them to be. All of the *Fleet* personnel.. There had been absolutely no Marines among the naval personnel waiting to surrender.

Chowdhury's mission had been to take the asteroid base, and obtain whatever evidence remained of who and what had been involved in the attempt to waylay *Pathfinder* and *Vanguard*.

The officer who met her to surrender the base formally, a DaGaman naval officer who identified himself as Lieutenant(jg) Fazendeiro, informed her they had not been resupplied for many weeks, since their two assault shuttles had headed off to secure *Vanguard* and had never returned. They knew that supply pods had been sent through the jump point for them, but they had no ship to retrieve them.

She hid a vicious smile at the knowledge that neither shuttle had returned. It appeared the little present she had left behind for Rockhead had worked after all. She wasn't going to miss him in the least.

Apart from the personal satisfaction that knowledge gave her, the information they had gathered from the base staff so far was not of much value. Lt. Fazendeiro was not forthcoming about much else. From what she could tell, most had no idea of the overall scope of their mission. Those who had possessed that knowledge had been lost with the shuttles.

Hopefully, they would still be able to gather enough intel on what was happening here to bring the DaGama Family to justice. So far they had not come up with anything worthwhile. Their presence in a Warner–owned system was enough to cause the DaGama family some serious problems, even if Chowdhury could not find documentary evidence to prove a connection to the attempted piracy of the Warner ships.

She was standing now in the main control room of the base, observing monitors as her people continued to round up stragglers. The DaGaman officers had given up their access codes and keys readily enough. Anyone who didn't was told they would not be fed. She called up a base schematic using Fazendeiro's code, and noted that they did not have vid feeds for every section. After some work with the system, she was able to bring one of the dead zone cameras online briefly.

A flash of movement on that monitor caught her attention. Chowdhury glanced to the right and saw that her second-in-command, Captain Brian Nash, was checking another set of feeds at the other duty station. He was a bull of a man, easily two meters tall, and massing nearly 120 kg, with arms and neck that dwarfed most people's legs. Not much of his face was visible to her behind his helmet and his heavy battle armor, but his stance and intensity were causing his Marines to unconsciously give him a wide berth.

His size caused some people to make the mistake of thinking he was slow. He wasn't. Not mentally, and certainly not physically. He generally had a calm disposition, despite his normally grim visage, with a great sarcastic wit, but he was absolutely merciless in a fight. Her jarheads had learned in their training that he was not someone you wanted as a sparring partner unless you were either a master in hand-to-hand or a masochist. That meant that he and Chowdhury were left to spar with each other, since most people held that if you sparred with her, you needed to be both.

"Captain Nash, you have supervision of the mop up. I am taking Fire Teams One and Two, First Squad; there is something I want to check out in that dead zone I have called up at the main station," Chowdhury said as she slapped her lid back into place and moved quickly toward the door, making a hand signal to get the eight chosen troops moving on her heels.

"Yes, ma'am," Nash's deep voice said over the com. "All right, what's the count, Sarge?" he asked the Marine on his right as he moved to where Major Chowdhury had been standing.

Chowdhury and her team headed up the corridors, deeper into the base. She rounded a corner, past the last working camera she had seen from the control room. At that point, she slowed their progress, providing fire cover and leap frogging down the corridor. Within 200 meters, the corridor took a sharp right, but there were alcoves to the left and then interspersed down the next corridor run.

She held up her fist to stop their progress. If she were going to set an ambush, this is where she'd do it. Great cover to fire from, and multiple fallback alcoves, some presumably with exits easily at their rear. Sheli hand signaled for Corporal Fitz Barker and PFC Johnny Tanzo to take cover positions and for his fire team mates to leapfrog over to the alcove on the left that looked to be empty. She signaled for them to check for explosives and expressed caution with her downturned left palm slowly moving to the floor at her left knee where she knelt at the corner.

She motioned to the next team and Gunny Barclay Medina and PFC Kevin Franks made their move to the alcove. She had heard a high pitched whine begin at the far end of the corridor.

"Everyone down! Take cover!"

The whine became the rapid fire of a pulse gun as it lit up the entire corridor.

Gunny Medina rolled clear, flattening into the alcove, but Franks was hit high on his left side. The impact on his armored shoulder spun him around and a flash of plasma-burned metal almost blinded them all and made their helmet speakers dampen to their lowest setting. Franks hit the ground facing away from where he'd been headed.

There was heavy fire covering the whole corridor now, with her Marines returning fire from their corners. Master Sergeant Steven Burik dashed out of his cover behind her and hauled Franks back to safety. Just as he came across the engagement horizon, he took a salvo to his back. A quick glance showed that the armor was severely damaged and he was hurt, but was at least slightly mobile. Similarly, Franks was in pain and down, but not unconscious. Both were behind her now. Barker and Tanzo were laying down efficient cover fire, as was Medina, but about every four or five seconds, the heavy pulse gun would open up again and they all had to dive back to cover.

Scrap, Chowdhury thought, as she reassessed the situation. She was pretty sure she had found the missing DaGama Marines. She thought that the heavy pulse fire was winding down again and was about to make a move for the alcove with Medina, taking part of her group with her, when the wall next to them exploded, toppling her team and sending her sprawling out into the corridor.

Major Chowdhury saw stars, and instinctively flexed muscles as she had been trained to do to force heavier blood flow to her brain. The clouds of smoke and dust were being illuminated above her as plasma blasts streaked overhead. She was in the middle of the corridor lying on her side. She was completely exposed to the enemy. She forced herself up and had to struggle briefly against a large steel plate on her right side., the light armor did not provide any additional power assist.

Her right hand didn't have her pistol in it any more. Her ears were ringing slightly, and the steady thumping sound of her blood in her ears reminded her to control her pulse. *Scrap,* she thought again as she could see what she thought were armored DaGamans moving quickly in her direction down the corridor. A quick glance to her sides didn't yield any clues as to the weapon's location.

Behind her, her men were clearing themselves from the rubble from a section of wall that had been destroyed. There was no catastrophic venting of atmosphere, so clearly this corridor was contained within a larger superstructure. They must have lobbed a grenade, or had planted some explosives ahead of her arrival. Her fire teams didn't appear to be in immediate danger as they pulled free of the rubble, but clearly she was, as a plasma pulse scored the floor in front of her. She drew her rifle from the container on her back, thumbed it hot, and began returning fire at the incoming enemy Marines. She started moving toward the alcove that still seemed to contain Medina, or someone who was shooting at the enemy.

As she settled into high cover over Medina, she finally got a clearer look down the corridor. There must be a full platoon of DaGamans advancing in detail toward them. The ones in the front ranks had heavy armor.

"Nash," she said calmly into the com.

"Ma'am," came the reply quickly.

"We're going to need some help down here. Soon. I think we uncovered the missing Marines," she said while firing off another burst from her rifle.

"I'm zeroing in on your beacon now, should be there in a minute or so," Nash's basso voice said with a rumble.

Like any good Marine Captain, he had apparently anticipated his Major's needs and was already on the move when she had commed him. She was counting on him bringing significant firepower to the situation. He and his reserve fire teams were heavily armored and ought to be sporting their heavy weapons.

Corporal Kazi Jones and PFC Shrongradov were moving to support in laying down suppressive fire on the DaGamans while the remaining team helped those buried under the chunks of steel and composite debris. Unfortunately, Sheli didn't have the luxury at the moment to worry about the men still down. There would be time for that later.

The DaGamans were within 50 meters. There were at least 20, maybe more. There was one of them down at least. Not a mean feat considering her Marines were lightly-armed and armored, while the enemy were both better armed and protected.

Suppressive fire from both sides crisscrossed the space between the groups now. Chowdhury pulled a pair of crashers from her kit, and lobbed one up at the closest group with her right hand – only to have it shot out of the air with an incredible explosion of light, sound and pressure waves. The one she had side-armed from her left hand ricocheted around and landed just in front of the alcove containing a couple of the heavily armored front line. It went off right in front of one of them as he stepped out to open up with his heavy repeater. The blast tossed him like a rag doll back into the alcove, slamming his armored form into his backup.

Chowdhury wanted to press the moment, but a massive salvo of heavy repeater plasma blasts streaked down from the DaGamans in answer.

The wider hallway was so thick with smoke, dust, debris and blast bolts that it was almost impossible to make out who was where. She was glad that she was armored and had self-contained air systems.

From down the originating corridor, Sheli heard the heavy thumping of Nash and what must have been at least half a platoon.

As Nash himself approached, he commed her, "Ma'am, where do I put the holes?"

"Downrange, Marine," she said, deadpan.

The coms erupted in some quick barks of laughter and then clear instructions from Nash and Chowdhury as they repositioned the reinforcements.

"Alright, I think we can handle that."

Two intrepid Marines set about positioning a heavy cannon. Nash and his immediate fire support were laying down clouds of heavy suppressive fire. The return fire was definitely lessening.

As the cannon opened up on the enemy, the destruction to the corridor was horrendous. Whole sections of walls, doors, ceilings, everything in its path was being shredded and flung around like it was confetti at a parade.

After a minute of the heavy barrage, Major Chowdhury called to her Marines to check their fire.

As soon as the firing stopped, Nash addressed the enemy.

"You have nine seconds to throw down your weapons and surrender unconditionally, or we will shred the walls around you and space you all."

"Eight," he paused, "Seven," he said again, not truly waiting an entire second. He made it all the way to four before a few guns started getting tossed into the smoke filled corridor.

The act of surrender seemed to ease the tension of the moment. Some of her fire teams started to move slowly forward, towards the small pile of arms. Before Chowdhury could warn them or call them back to their position, the air erupted with heat and destruction. Along with the guns, several grenades had been tossed out. The successive explosions were deafening and threw the three Warner Marines back through the air. Cracks could be heard in some outer base hull walls and atmosphere began to whistle out, wind picking up debris and shrapnel alike, shifting it airborne and out the top of the hallway. She wanted to scream. They knew better than that! What did they think they were doing? She said none of that. Recriminations could come later.

"Return fire," Chowdhury called into her com, a fraction of a second before every Warner Marine with secure footing opened up. Magnetic boot locks clamped down all around her as her men and women "dug in" against the steady loss of pressure.

The originating corridor that led back to the control room began to seal off against the depressurization.

"Ma'am," Nash almost shouted against the background noise.

"They made their choice, Nash, give them a quick death," she said coldly.

"Yes ma'am," came the resigned reply.

"Gunny, get the wounded and anyone without a pressurized suit behind that blast door immediately," she called to Gunnery Sergeant Spire on the other side of the hallway.

"You heard the major, all of you without pressure that can move, do it, the rest of you over here, help get the wounded on the other side. You two," he said, pointing at two privates, "will stand guard with them "

A few minutes later, they stopped firing. There was almost nothing left of the hallway. Open space was visible above them. There didn't appear to be anywhere that the DaGamans could be hiding.

Without moving from her firing position, Chowdhury called out over the com. Her eyes continuing to scan for new targets

"Fire Teams six and eight, careful forward, make sure there is no one else out there. Control, we'll need you to identify an airlock near us. We have wounded, about 500 meters on the other side of your camera vision." she commed.

"Major, there is one about a half a kilometer away, that seems to be the best option," came the reply. "Help was already on its way, including corpsmen. They're with your wounded now."

The Warner Marines calmly swept through the debris field checking for stragglers or survivors, while the quiet of open space seemed to enfold them.

Captain Nash and his repeating rifle followed closely with the second team, and she followed him. As they neared the far end of the corridor, they found what appeared to be a staging area, and offices. She gave orders for teams of two to carefully investigate all the rooms. She didn't want any more casualties among her Marines.

Corporal Jones commed to let her know he had found something she'd want to see. She followed his beacon to his location.

She stopped and entered the office he occupied with Nash at her back. It had been the office of the base commander, a Commander D'Agostinho, based on the nameplate. The team had already cleared the room and were investigating. If they were going to find hard evidence of the purpose of this base, here would be the best place she could think of. While there was no atmosphere here, most of the room was undisturbed as a function of the slower loss of pressure it had experienced.

She moved over to the commander's data terminal. As her men began going through the drawers and cabinets, Chowdhury checked the computer system. Power was out. *Blast.* She checked the system connections, and called over one of the combat engineers.

After a few minutes of waiting for his prognosis, he shook his head while still under the system. "No good ma'am, can't get power to it."

"Are there any local storage systems connected to it? It looks like he maybe had this one off main network, right?" Chowdhury asked quickly.

"Yes ma'am," he said, as he came out from under the system holding two solid storage units. "Most likely they'll be encrypted, maybe even using a system key, so I'll grab everything here. Hopefully they weren't fried or wiped when power went,"

"We'll see then," she said while reviewing the rest of the room.

The rest of her team seemed to be coming up empty.

They were soon cycling through the airlock that Control had indicated to them, and returning to the rest of the DaGaman prisoners along with the rest of her dragoon.

Her team had made copies of all of the security footage available, but there didn't appear to be much. All of it was set to write over as needed, and there wasn't much more than thirty days' worth available. The likelihood of finding anything useful was slim, but they would sort through it anyway. The best opportunity lay with the drives.

She needed to check the drives quickly to see if there was anything to recover. She had one of the terminals passed to her and connected one of the two drives. As the device tried to access the drive, it became obvious that it was, in fact, encrypted. The test of the second drive yielded the same. None of the standard cracking tools she had with her were of any use; it was actual mil-grade encryption. This was not going to crack easily.

"Lance Corporal Simms," she said turning to the short young man to her right, "get me a link to *Dagger* immediately. We will need to get these drives to the techs onboard as soon as we can."

"Yes Major, com link being established," he replied in an oddly juvenile alto-tenor voice.

As the com link was coming online, alarms all over the control room began sounding, along with flashing lights.

"What is that?" she asked everyone in range.

Her teams were checking every panel and board around.

"It appears we have another breach in the base, ma'am," one of her young female officers responded from behind her. Lt. (jg) Belatroix by the sound of it.

"Show me where," Chowdhury said turning.

They were in for a more challenging evening that she had thought. Nothing ever comes easy.

CHAPTER 15

Dagger

29 August

Captain Fyonna Johnson approached the hatch with trepidation. She had wrestled with this dilemma for several days and had finally run out of options, without getting any nearer to a solution. Even her early morning walk around the ship had failed to allow her thoughts to crystallize. She knew that, as *Dagger's* captain, this was her responsibility, and she did not want to add to the burden the commodore was already carrying, but she had no clue where else she could go for advice. She heard the chimes announcing the end of second watch as she rounded the last corner before her destination. She almost returned to her cabin when she realized the lateness of the hour. *No, this needs to be resolved now*, she thought. She arrived at the hatch and straightened her uniform unconsciously while she set in her mind the things she wanted to say. Finally, she touched the admittance chime to announce her presence. Within a few moments the hatch slid open to show the tall, fiery-haired commodore. His uniform and grooming were impeccable, despite the hour being just past midnight. Captain Johnson had expected no less. She still was not sure if this showed a personal vanity at work or if he simply could not allow anyone to see him at less than his best. Or was he trying to put forward the best face possible because he believed that to be his duty as a commanding officer? She personally leaned towards the latter explanation, but she still wasn't convinced one way or the other.

"Captain Johnson how may I help you?"

"There are some things I would like to discuss with you, sir, if I am not imposing."

He stepped back to clear the doorway and motioned her inside.

"Not at all, Ms. Johnson. Please, come in. What is it you need to discuss?"

The change in honorific actually helped her feel a bit more at ease. Most commodores or admirals would use their flag captain's given name in a private conference, but it was impossible to imagine *this* commodore unbending that much. The use of 'Ms.' however, a title that had been hers ever since she graduated the academy, was a comfortable one, and one she knew she deserved. 'Captain' was an entirely different matter.

"Well, sir," she began as she took the offered chair and he sat opposite her, "I'm not sure where to begin." She looked up at him and, at his nod, continued. "Frankly, I'm a little embarrassed to have to bring my problems to you, but I have been struggling with the prospect of unifying the crew and getting them to work together. The original *Dagger* crew is resentful of the situation which has robbed them of their earned leave while at Earth, and the *Vanguard* representatives feel like they are the minority who has to gang together against the remainder of the crew. The new crew that was not a part of the *Vanguard* crew is leery of both groups. There are at least two camps starting to entrench with each other as the designated enemy instead of any DaGama thugs or *Pathfinder* pirates. I need to find a way to bring them together."

She took a deep breath before concluding, "You seemed to have a great amount of success in that regard aboard *Vanguard*, so I thought you might be able to at least point me in the right direction."

Commodore Brighton was silent for so long that she began to believe she had overstepped her place, after all, but just when she was preparing to apologize and leave the cabin, he said, "Thank you for the compliment, however undeserved. I am flattered that you would come to me with your concerns. This is, quite possibly, out of my area of expertise as well. With the possible exception of *Vanguard*, I have never had any great success in creating cohesiveness within the crews that I have commanded. Obviously, I have thought about every action and decision that was a part of that journey and second guessed many of the decisions I made, but the fact remains that without a concerted effort from all of the officers and crew of that launch, we would not have survived the ordeal."

"Sir," she objected, "that is my point. You molded the crew into what you needed to have in order to be able to accomplish the mission. I don't know how to do that."

"Captain, I think that you attribute too much credit to me. No captain can mold a crew without its consent. Only tyrants can impose their will on a crew and force a result. I don't believe this to be your objective here," he made this statement into a question with his piercing look and a raised eyebrow.

"No, sir, that isn't my goal, but I feel that you underestimate your role in the journey."

"No, Captain, I don't feel that I do," he said after a slight pause. "I provided the drive and the direction, but the crew made up their minds to put in the effort," he said and looked off into the distance, his eyes taking on a slightly unfocussed look as if he were picturing the scene in his mind. "Did you ever know Captain Mulvay?" he asked suddenly.

"No, sir, he had retired before I completed my time at the academy," she said slightly taken off guard by the non sequitur.

"I was just remembering something that he told me once," he began. "He quoted to me from the Bible. It was the only time that he did so that I can recall and it has taken me nearly twenty years to understand the lesson that he was trying to teach me then. He said 'Where there is no vision, the people perish.' "

"What did he mean by that, sir."

"I thought at the time that he was telling me to plan ahead, but I've come to believe that he meant not only that, but also that we need to give the people under us something to believe in. A vision of what should be. A goal to strive toward, both in physical reality and in character."

"So I need to have a common goal for all members of the crew to work toward in order to draw them into a cohesive unit?"

"I believe that is part of the answer to your dilemma, but I think there is more to it than that, Captain."

Johnson sat for a few moments while she absorbed the new thought the commodore had just introduced. It was nothing new, simply a matter of looking at an old concept in a new way. "Thank you for your time, Commodore," she said as she rose to leave. "You have given me a starting point and a few more things to think on."

"You are welcome at any time, Captain," he said as he stood and accompanied her to the hatch.

Johnson continued to think on these things as she made her way back to her quarters. She had turned into the forward cross corridor when she abruptly turned back to the quarters she had just passed. Her XO had just finished his watch and should be in his quarters, unless he had other business to attend to.

She touched the chime and again waited for an answer.

The hatch slid back to reveal Lieutenant James Grant standing framed in the opening. If there was a physical opposite to the commodore, then Jim Grant was as close as it was possible to get. He was barely 160 cm in height and weighed nearly 100 kilos. His round face wore its usual smile and he bubbled – *gushed?*- with enthusiasm. He was always ready to tackle any problem, even at the end of his normal duty shift.

His head was cocked to one side and he motioned the captain into his quarters. Just as with the Commodore's berth, and her own, the front room of the XO's quarters functioned as his office when not on watch. He motioned her into the visitor's chair and he perched himself on the protruding edge of the desk built into one wall.

"How may I be of service, Captain?" he asked with his usual cheerfulness.

"I've made a decision that I need you to implement, Lieutenant," she began.

"Yes, ma'am, what form is this implementation to take?"

"You are aware of the discipline issues that we have had in the crew spaces the last few days?" she asked.

"Of course, ma'am," he said, suddenly wary. "I have dealt with most of the issues and have only forwarded those cases that warrant a Captain's Hearing," he added and started to rise to emphasize his case. He was careful not to mention the fight in the boat bay which she still needed to resolve.

"I have no problems with the way you have handled anything, Jim," she said to calm him and motioned him back down into his seat. "Rather, I've been trying to think of a way to smooth the ruffled feathers and try to make a start at integrating the crew. We lost a lot of the original crew to that virus that beached Captain Andrus and the rest. We need to pull them together into one seamless unit instead of two or three armed camps," she said, knowing that she was again skirting very close to his personal territory as XO. Traditionally, the executive officer on any Fleet ship was responsible for setting up the drills and any other appropriate measures to weld the crew into a fighting unit. She didn't want to undercut his authority or take away duties that centuries of tradition said should be his. She watched his face to measure his reaction. His smile was still in place but it appeared to be slightly strained.

"I think you have done wonders, so far, but I feel this is an unusual situation. We need to give them a common enemy and a common goal. I will stand in as the common enemy until we can find something more appropriate. I think the random drills at all hours of the day and night that you have already implemented should take care of most of that, but I need you to create a project that will take all of their extra time. There needs to be an urgency to complete the project. Do you have any ideas?"

Grant sat still for a few moments then sprung up as if he suddenly realized that he was sitting on a beehive. He grabbed a folio from a cubby in the top of his desk and activated the holo. The diagram that floated out of the folio was immediately recognizable to Johnson.

"Where did you get that?" she called out as she sprung to her feet.

"It's just the engine specs from *Pathfinder*," he replied calmly with a look of confusion on his normally placid features.

"I know what it is," she said through clinched teeth, "I asked you where you got it."

"It was in the last download from Betre before we transited the JP."

"What?" she sat down in shock, "That should all still be classified."

82

"No, ma'am," he said finally understanding her sudden change of behavior. "This has the security stamp from Admiral Cosina's office. We're cleared for this," he assured her.

"I've been looking at the specs and I think we can duplicate this on *Dagger*," he stated calmly. "Would that be a big enough project for your needs?"

Captain Johnson calculated the space and area needed to build the capacitors large enough to move a ship of Dagger's mass and turned to her XO. "It's definitely big enough, but I don't think you can make it work. There isn't room to build capacitors large enough to do the job."

"I think a parallel coupling of smaller units will work for the capacitors instead of the larger, more massive installations they used here," he said, pointing out the relevant parts in the diagram.

"Okay, but here are your limits," she said cutting him off before he could truly warm to the subject. "At no time will you disable either the existing engines or any of the weapons systems without my express approval, and you will take Master Chief Mackey, Warrant Long, Chief Giovanni and Master Chief Durrant as your assistants," she said, naming the two leaders on each side of the crew stand-off.

"Yes, ma'am," he replied with a chuckle as he pictured assigning Long and Durrant to a common project. "I'll see to it, but it is a wicked thing to do."

"I do my best," she said with a grin and left his office to try to find her bed. "Have fun," she said as the hatch slid closed behind her.

"Oh, I will," he said to himself as he sat in the vacated chair and chuckled at the ceiling.

CHAPTER 16

Dagger
30 August

Chief Petty Officer Maria Giovanni came around the last corner, arriving just before the requested time, and chimed the hatch. She stood quietly, almost at full attention, awaiting the XO's response. A few moments later, Lieutenant James Grant was ushering her inside the small office.

"Nice to see you, Chief," Grant said as he pulled out and turned around his chair from the writing desk against the wall, "Why don't you take a seat?"

Giovanni sat in the proffered chair as the XO did the same in a chair he pulled in from the adjoining room. Once he was settled, he looked up at her, and offered a casual smile.

"Sir, I am here in order to comply with the instructions of the captain. I was asked to report on everything that happened during the incident in the boat bay," she said stiffly.

"All right, Chief. Why don't you start at the beginning of the incident as you recall it, please," the XO suggested as he readied his notes.

"Well, sir, I guess it all started when Warrant Long became belligerent with Ensign Keone and we all came over to back up the ensign," she began.

"Go on please, Chief," the XO encouraged as his eyes engaged with hers.

"Sir, it just escalated from there. Long was popping off and the ensign wouldn't back down. Bosun Mackey got involved right from the beginning, and it was him that threw the first punch," she stated.

"So you are saying that Bosun Mackey threw the first punch; that he is the one who escalated the confrontation to physical violence?" Lieutenant Grant said with incredulity in his voice.

"Yes, sir, that is exactly right. The bosun started the fight, sir," she declared.

"Tell me exactly how that transpired, Chief," Grant said with a slight frown.

"Well, sir, like I was saying before, Long was getting heated with Ensign Keone because the ensign wouldn't back down. Master Chief Mackey was getting between them but was right in the ensign's face. Long was threatening the ensign, I mean there were loads of us backing the ensign up by this point, so we thought he was just blustering," she paused for a moment to get a breath, "but when the ensign told him he shouldn't be getting involved with an officer, Bosun Mackey just stepped up and leveled the poor ensign," she managed to get out in her next breath.

"I see," said James, as he leaned back and peered at her from under furrowed brows. "And so, Chief, you are saying that Mackey knocked the ensign down?"

"Well, sir, he about killed him with one blow," she continued, getting very animated about her description. "The bosun fired off with one of those ham-hock fists of his and crushed Ensign Keone's face in – knocked out at least two teeth and the ensign was out cold before he hit the deck," she replied as if she couldn't believe he didn't already know this.

"And I presume that the rest of the scuffle that occurred was completely in reaction to this event?" Grant asked her, covering over his notes as he leaned back in across the desk slightly.

"Yes, sir, that was what started the fight," she replied matter-of-factly.

"So are you purporting then, Chief, that neither Ensign Keone, nor anyone 'backing him up' as you put it, escalated this confrontation in any fashion beyond the provocations offered by Long and Mackey?" the XO queried her as he raised an eyebrow and leaned in even closer.

"No, sir, well... there may have been some jibes or words tossed out, but I don't think any of us thought the situation would escalate to violence," she replied, looking only slightly apologetic.

"Chief, you say that you feel that none of you thought it would escalate to violence then, correct?" he asked in a conversational tone.

"That's right, sir," she replied.

"So why did all of you feel the need to come over and 'back up' the ensign, if none of you were worried about any physical confrontation?" the XO asked with a look of genuine confusion on his face.

"Well sir, um, we didn't think it would get physical, but we wanted to show these grandstanders that we backed up our officers, no matter who was trying to get over on him with words," she said, looking for a path clear of the trap her own report threatened to close around her.

"So, are you saying, then, that you and your counterparts on the ensign's side did not feel that this particular officer, Ensign Keone, was capable of talking to two non-coms about the work to be done without your input? Please enlighten me further, Chief, because it almost seems that you are saying that the situation did not appear to be headed towards physical confrontation until after you and your group joined the conversation, is that accurate?" the XO asked in earnest.

A short pause hung in the air before Giovanni answered.

"Sir, I am not trying to get my words mixed up. I guess that it is possible that the rest of us getting involved made Long and Mackey feel like they had to be macho and step up their bluster. I would say in thinking back, we probably didn't help the situation like we wanted to. I still didn't think they would dare start a physical confrontation, certainly not when it was clear it would be like twenty on four, but I suppose it is possible that it contributed to their escalating the confrontation," came her long-winded response.

The XO looked at her for a moment, then down at his pad to make a couple of quick comments there before looking back up at her and saying, "Is there anything else you wished to add, Chief Giovanni?"

"No sir, I think I covered it," she replied.

Lieutenant James Grant looked down at his notes and made a couple of quick marks, and then looked directly at her and said, "Very well, that will be all, Chief. You're dismissed."

Maria Giovanni nodded and stood. She came to attention briefly before Grant nodded and waved her out. She about-faced and headed out into the corridor again. She walked quickly away from the XO's office and was not at all surprised to find Durrant waiting around the next corner in the corridor.

"What did he say? What did you tell him?" Durrant blurted out before Giovanni could say anything at all.

Giovanni gave Durrant the run-down of the conversation, pointed out that she had clearly placed the blame on Long and Mackey, and that she had definitely emphasized that Mackey had thrown the first punch that started it all.

"You realize that the captain is not going to have any choice but to hold a court martial for Mackey now. That self-righteous oaf is going to wind up in the brig for years at the very least," he said with a wicked grin on his face.

As Giovanni smiled back at him and they started down the corridor together, he looked over at her, grinning even more broadly, and said "I'll be back in the bosun's berth before the week ends."

CHAPTER 17
Dagger
30 August

The assault shuttle settled into a smooth dock with *Dagger*. With the near constant comings and goings of the shuttle, the exterior dock made more sense than an internal landing ,with its pump down and restoration of atmosphere to the boat bay. WO Truman Hijironoma flipped the all clear signal to the on position, indicating clean atmosphere in the sealed dock at the port airlock. Major Chowdhury motioned for her Marines to follow her out, and she headed to the airlock. She regretted having to leave so many of her Marines on Le Vesconte to guard the DaGaman prisoners and she felt the loss of Captain Nash, her XO, almost as much. The Marines in Bravo Platoon, her only remaining platoon, had already been instructed to assemble in one hour for debriefing. During that time, she needed to meet with Commodore Brighton and go over what had been learned before they could plan their next moves, and an hour would give the Marines time to stow their gear and get cleaned up before reporting to the debrief.

Sheli went to her quarters. She tossed her gear onto her bunk, aware that neither would see any use in the near future.

Apart from the forty-four Marines she had left on Le Vesconte, she had the rest of her Panther Dragoon, which consisted of forty-two Marines under Lieutenant Shaun James, Bravo Platoon's CO, and now her acting XO with Nash gone. She also had at her disposal, if urgently needed, the fourteen Marines aboard *Dagger* as the security contingent.

Since *Dagger*'s naval crew did not exceed 150, there were fourteen Marines in the ship's security team, commanded by Lieutenant Vitek Gvozdzius. As a security team CO, he was not in Chowdhury's normal chain of command, but if needed, his team was an asset that could be called upon. Chowdhury quickly accessed the inventory and personnel files she wanted to check. She gathered the two data chips she had filled and left her quarters, after less than ten minutes. She made her way to Commodore Brighton's Ready Office, which was adjacent to his quarters and near the bridge.

She knocked once on the door. A faint "Enter" was heard over the whoosh of the opening doors.

"Commodore Brighton, Major Chowdhury reporting for mission debrief, sir," she said coming to attention in the doorway and snapping a crisp salute.

He was already standing up from behind his desk. He quickly returned the salute and said, "At ease, Major. Tell me what you've found beyond the report you sent in."

The next twenty minutes passed with Major Chowdhury reporting on significant events that had transpired on the asteroid base, her decision to move the prisoners to Le Vesconte and leave Captain Nash with his Marines there to guard them. She answered the anticipated questions, and concurred with his thoughts on where to go from here.

"We have yet to receive contact from *Yargus*, although I hadn't really expected it yet," he said to her.

"We're going to follow her, then?" Chowdhury asked, knowing the answer.

"Yes, Major, we are. I am sure Nash and his Marines will be sufficient to their assigned task, and I know you and Bravo Platoon will have your hands full once we manage to overtake *Pathfinder*. Do you feel they will be sufficient to *your* task?"

"Yes, sir. Absolutely."

"Very well. I will trust your assessment. Will you walk with me to the bridge until you need to leave for your team briefing?" he asked as he stood, blocking the chrono on the wall behind him. "You have what, another twenty-five minutes before you need to go meet your Dragoon?"

"Yes, sir, I have about twenty-five minutes before I need to be there," she said without any indication that she had needed to check the blocked chrono, "I can walk with you as you go, sir." she replied, standing to follow him to the door.

"Thank you, there are a few more things to discuss," Brighton made the statement an invitation by inflection, but as always, there was no missing the implied command in his words.

"Of course, sir," Sheli replied.

A few moments later, they were passing through the bridge doors.

"Comm, send word to *Avram* and *Foundation*, we will be moving out shortly," the commodore said as he moved to the plotting table with the major. "Let's go over some of what we may see, Major. I want to have plans for all eventualities, and I would appreciate your strategic thinking on what I have laid out here," he said to Chowdhury, indicating the array of diagrams, star-maps, and scenario-loaded chips on the plotting table.

Chowdhury nodded and sat down to begin assimilating what he had been working on.

"Once again, I apologize for having co-opted part of your bridge, Lieutenant," Brighton said to James Grant who was Officer of the Deck and at the command chair.

"Sir, there is no need to apologize. I am always glad to have you on the bridge," came his quick, and surprisingly sincere, reply.

"Astrogation, are you ready to set course to follow *Yargus*?" Grant turned to the Astrogation Officer and asked the question almost as soon as Commodore Brighton had sat down.

"I am, I mean, yes, sir," Lieutenant(jg) Marcus Mythrodakis answered with all the confidence he could muster, clearly perturbed by having the commodore on the bridge.

"Let's hear the plot then, Astrogation," the XO replied.

Mythrodakis called it out, and after a nod from the XO, Navigation set the course, called the course confirmation back and Comm sent it to *Avram* and *Foundation* so that they could log the course change in their "fleet separation" log. *Avram* and *Foundation* would not be making the course corrections themselves, since they would be staying behind.

"Course set, Lieutenant," came the call from Astrogation.

"Course received by flotilla companions for log, XO," Comm responded on the heels of Astrogation.

Before Chowdhury could begin to make the first of her comments to Brighton on his plans, Lieutenant Grant turned to him from the command seat and said, "Commodore, with your permission?" indicating he was going to proceed with the course change. He didn't have to ask, of course. As OOD, he was responsible for the bridge and had received orders to proceed, but it was a show of respect to ask any flag officers for permission to proceed when they were on the bridge. She was glad Grant had considered that in respect of Brighton.

"By all means, Lieutenant," Brighton responded and turned back. Brighton focused his attention on Major Chowdhury as *Dagger* slowly began changing vector to the new heading.

CHAPTER 18
Dagger
31 August

The hatch to the meeting room slid silently open on its track, but clicked as it locked into the open position. Twenty-three men and women marched smartly through the opening, dressed in their formal uniforms. Each made his or her way across the room to stand in front of the table until there was no further space, then turned and faced the seven officers seated behind the table and drew to attention. Their faces attempted to remain impassive, but occasional nervous or furtive glances marred the uniformity of the display. Four armed Marines followed them in and positioned themselves so as to block anyone who attempted to enter or leave without permission.

As previously, the captain did not immediately speak, and the tension in the room bore down on the guilty consciences arrayed before her.

After an endless minute, the executive officer, Lieutenant Commander Grant, rose from his seat to the right of the captain. He reached to the center of the table to retrieve a wooden hammer that sat next to a small brass bell suspended from a stand. Several of the individuals visibly tensed even further. When he made no move to strike the bell, the same crewmen relaxed a bit.

"It is the decision of Captain Johnson," Grant began, his eyes seeking and locking momentarily with each of those opposite him, "that this incident be resolved at a Captain's Mast. The captain will determine punishment for each person involved at her own discretion. There will be no legal representation, no mandatory penalty, and no appeal until the punishment has been completed.

"If any of you would prefer, it is your right to stand before a court martial, be represented by legal counsel, and face the defined penalties of the Warner Naval Code. If such is your choice, you will be held in custody until this ship returns to a naval base where such a court may be convened."

He paused to make sure the gravity of the situation had a chance to sink in, then continued, "Any who wish to seek a trial by military court rather than accept the captain's judgment, take one step forward."

The wait seemed even longer in the silence, but no one moved to accept the second option. It was not surprising. Their current deployment did not have a scheduled end, and being confined in the brig for an indeterminate time certainly held no appeal. When enough time had passed, Grant returned the gavel to its place without striking the bell and seated himself.

Captain Johnson rose in her place, her jet black dress uniform ill-fitting and baggy due to the hardships she and her *Vanguard* crewmates had so recently undergone. There was nothing frail or weak-looking in her stern countenance, however.

"Ensign Polunu Keone," the captain's strong contralto voice called out, "stand forth."

Keone took one step toward the row of tables before him. He had been the first to enter the room, so he was far to the captain's left and she turned in his direction as she addressed him.

"Ensign Keone, after interviewing those involved and reviewing the available evidence, you are charged with dereliction of duty and conduct unbecoming an officer. Do you have anything to say in your own defense?"

Polunu did not hesitate in his response, but he could not bring himself to face the captain, directing his words at the far wall straight in front of him. "No, ma'am. I have no defense for my actions."

"Very well, Ensign. You will receive a formal reprimand, which will appear in your permanent record and will include a guilty finding. You will forfeit all accrued seniority, and today's date will be inserted in your personnel jacket as the date of promotion to the rank of ensign. In addition, you will undertake one thousand hours of additional duty on a project to be named hereafter. Do you understand the terms of your punishment?"

"Yes, ma'am," he stated, saluted, and then returned to his place in the line when she returned it.

"Master Chief Petty Officer Derrick Mackey, stand forth."

Mackey took a step forward and waited, still gazing straight ahead.

"Master Chief, you are charged with striking an officer, assault, and disorderly conduct. Do you have anything to say in your defense which might mitigate the punishment you are about to receive?"

"No, Captain," his deep voice said quietly.

"Very well, you a—"

"Captain, if I may?" Keone interjected in a firm voice, still not looking directly at her.

Johnson looked at Grant briefly, but his look of puzzlement told her that he had no more notion of what the ensign intended than she did. Turning back, she nodded once and the man advanced one step, then continued.

"I would like to speak in the bosun's defense, ma'am. In my opinion, Master Chief Mackey was provoked into taking the action that he did. As the officer involved, I request that the charge of striking an officer be dropped, ma'am."

Having said his piece, Keone returned to ranks. Johnson was surprised at this request, but she maintained an impassive aspect. Not so Durrant. A look of shock quickly turned to one of anger which he made no effort to conceal.

"Thank you, Ensign," the captain said after a moment's consideration, "but your request is denied. While you are the one who received the damage at Mackey's hand, his transgression was not solely against you, but also a violation of military law. As such, it is not within your power, nor within mine, to simply dismiss the charge."

Keone's face did not react to the rebuff in any way, nor did Mackey give any indication that his fate had just been dealt a serious blow. Durrant's glower, on the other hand, had completely disappeared, though he seemed to be making some effort at disguising his grin.

"Master Chief Petty Officer Derrick Mackey," Captain Johnson again addressed her massive friend formally, "your many years of extraordinary service have been taken into account in your favor. Still, there is no question that you are guilty of the charges. Therefore, you are hereby reduced in rank two grades, to the rank of Chief Petty Officer, with no accrued seniority. You are stripped of the title of Ship's Bosun. You will be docked 45 days' pay, a sum which will exactly cover the medical expenses incurred on the day of the fighting. In addition, you are assigned one thousand hours of additional duty on a project to be named hereafter. Do you understand the terms of your punishment?"

"Yes, ma'am. Thank you, Captain."

Johnson felt sick at heart to hear Derrick thanking her for the punishment he had received. Unfortunately, there was too much evidence against him to find him not guilty, and she was pushing her Captain's Prerogative to the limits, and perhaps beyond, by not drumming him out of the service and throwing him in irons for the next twenty years. It was still possible, though she judged it unlikely, that her discipline would be overruled by the Warner Naval Board when they returned to Earth. She would face that if it came, and that was less important to her personally than being able to face herself in the mirror each day knowing that she had done her duty.

The captain noticed as she turned to face the next man on the list that Durrant couldn't contain his euphoria, now that he was again the ranking noncom on the ship. *Just wait, mister,* she thought. *Your turn is yet to come.*

"Warrant Officer 2 Stephen Long, stand forth," Johnson called out. The man advanced one pace, then resumed his stiff stance and unfocused stare.

"Warrant Long, you are charged with disorderly conduct and inciting others to disobey orders. Do you have anything to say in your defense?"

"No, Captain," he said.

"Very well, I find you guilty of both charges. You are hereby reduced in rank one grade, with no accrued seniority. You will also contribute one thousand hours of extra duty to a project to be named at a later time. Do you understand the extent of the punishment you have received?"

"I believe I do, Captain, but I have a question or two."

The captain paused, not sure for a moment what to do with this departure from the norm. "Proceed," she directed.

"First, this project," he began in a conversational tone that put the captain's teeth on edge, "will we hear the details today, or does 'at a later time' mean several days or weeks from now?

"Second, is everyone being assigned the same project, or are there a number of projects being started? And third, is there a time limit on completing the project or projects?"

"Well, Warrant, you seem to be thinking ahead," Johnson responded, "for once. So here are your answers, for planning purposes: before you leave this room, the same, and yes. Satisfied?"

"Uh, if I might ask an additional question—"

"No, you may not," the captain answered at once. "Back in ranks, Mister."

"Aye-aye, Captain," Long acknowledged at once, gave a moderate imitation of a salute, and moved hastily to comply.

"Chief Petty Officer Maria Giovanni, stand forth," Fyonna directed. When she had advanced herself, Johnson continued, "Chief Giovanni, you are accused of disorderly conduct and inciting others to disobey orders. Do you have anything to say in your defense?"

"No, ma'am."

"I find sufficient evidence to prove you guilty on both counts. In view of your cooperation in our investigation, you will not be reduced in rank. You will, however, forfeit your accrued seniority and are assigned one thousand hours extra duty on a project to be named hereafter. Do you understand the nature of your punishment?"

"Yes, ma'am," the slim CPO said and returned to her former position in the line after a quick salute.

"Master Chief Petty Officer Josiah Durrant, stand forth," the captain directed next. He did so, and the grin on his face was plain to see. It set Fyonna's teeth on edge even more than Long's casual speech had.

"Master Chief Durrant, you are accused of disorderly conduct, insubordination, and inciting to riot. Do you have anything to say in your defense?"

The difference in the charges was not lost on the non-com, and all traces of a smile disappeared from his face. "Captain, I was not insubordinate, nor was I inciting others to riot," he said simply.

"Oh, no?" the captain rejoined. "Were you not trying to urge others to fight with their shipmates?"

"Captain, I..." he began, then cleared his throat and started again. "Captain, there was already a dispute between Long and the ensign before I arrived. I was simply standing by to support the ensign in discharging his duties, ma'am. I said nothing to the ensign of an insubordinate nature."

"I see," Johnson said. She picked up a sheet from the table and referred to it while she spoke. "According to testimony received and recorded in the log of this Captain's Mast, you said that Commodore Brighton was 'twelve different kinds of idiot' and that he was a 'swaggering, overbearing, tin-plated dictator with delusions of adequacy.' The testimony further states that it was this statement which directly led to the fighting. Several of those involved say that you called or motioned them to come and join what was a private matter between Ensign Keone and Long and Mackey. That sounds like sufficient evidence to uphold all three charges." Her voice had no emotion in it at all, and Durrant swallowed to ease a suddenly dry throat.

"I never said that about the commodore, Captain. It was the ensign who said that."

"Not according to the testimony of others, Mr. Durrant."

"Who said so? I demand to be allowed to cross-examine them," he said, and panic was the dominant emotion to be seen in his eyes now.

"Denied," Johnson said flatly. "This is not a court martial, and you have waived that right by agreeing to a Captain's Mast. I will allow you, even now, to opt for that setting if you so wish, but I would warn you that the combined mandatory sentences for those offenses, if upheld, would be ten years, six months of incarceration, followed by dishonorable discharge from the Navy. Is it your wish that this matter be held over for the convening of a military court, Mr. Durrant?"

The big man had grown pale at the talk of incarceration, but his voice was clear and strong when he replied. "No, ma'am, that will not be necessary. I would like to deny again having made such a statement and explain that by gathering others to the scene I was attempting, through a show of ready force willing to stand with the ensign, to discourage Long and Mackey from taking any action. That I was not successful should not, after the fact, be construed as trying to incite such an outbreak."

"Ensign Keone, did you make the statement which I have attributed to Mr. Durrant?"

"Yes, Captain," he answered at once, "I made the second statement, but not the first. I will apologize to the commodore at the first opportunity, Captain."

Durrant grinned again, a mistake he should have known enough to avoid. "Very well, Master Chief, I will reduce the punishment I had intended to give you. You are hereby reduced in rank one grade, with no seniority. In addition, you will fulfill one thousand hours of extra duty on a project to be named later. Do you understand the punishment I have just described?"

"Yes, ma'am," he said, still grinning.

"Is something amusing you, Senior Chief?" the captain asked archly.

"No, ma'am. I'm simply pleased that the punishment was not more severe, as the captain reminded me it could have been. Will I be moving back to the bosun's quarters, ma'am? You know, since I'm the only senior chief on the ship."

Johnson grinned herself, now that she realized what Durrant had been thinking. "I'm afraid you're mistaken, Mr. Durrant. You are not the only Senior Chief on the ship. Senior Chief Quèneau has twelve days' seniority over you, and she will be taking over the bosun's duties."

Durrant reddened as he slunk wordlessly back into the line. No grin was in evidence now.

"Specialist 2 Keith Abney, Specialist 2 Claire Paul, Specialist 1 Pierre Fazan, Specialist 1 Angela Wetzker, Crewwoman 1c Sadie Vreeland, Crewwoman 1c Janine Van Der Wildt. Crewwoman 1c Pilar Trivedi, Crewman 2c Bill Sydow, Crewman 2c Dan Gregory, Crewwoman 2c Samantha Trajanovic, Crewman 2c Charles Corcoran, Crewman 3c Roberto Alcaraz, Crewwoman 3c Kjira Soderburg, Crewman 3c Jack Cuskeran, Crewman 3c Hein Bolon, Crewman 3c Denis DeGeus, Crewwoman 3c Diann Martin, Crewwoman 3c Brook Adibi stand forth." The captain again had to refer to her notes to complete the entire list.

"You are all charged with disorderly conduct, and disobeying a standing order. Do any of you have a statement which might mitigate the punishment you might receive?"

Captain Johnson waited, perhaps longer than necessary, and the stillness in the room remained undisturbed. "Very well, you are all assigned one thousand hours of extra duty on the same project you've already heard so much about. If any of you have any questions regarding this punishment, speak up." Again, she waited, and the same silence was the only response.

"Back in ranks then," she directed.

Once the uniformity of the line was restored, she looked each of the assembled men and women in the eyes, to be sure they were all focused on what she was about to say, rather than on what she had said already.

"Before I begin describing the project to which you will be devoting all of your off duty hours for the next few months, I need to tell you that some of you are about to be let in on what is probably the most closely held secret in the Warner Space Navy." She could see that she had everyone's undivided attention now.

"I say 'some of you' because those who served on *Pathfinder* and *Vanguard* already possess the information I am about to relate. And I say 'probably' because the secret I'm letting you in on was not perfectly kept, and that fact is the reason that you are on this ship now instead of on liberty. Be aware that this information is not to leave this ship, under penalty of treason charges."

The stillness of the room broke as those in line began looking at each other to see who knew what. The captain's voice brought their focus back at once.

"Warner Naval R&D has created and successfully tested a shipboard jump gate generator system. That system is installed on *Pathfinder*. That system is vital to Warner's survival as a viable corporation. That system must be retrieved."

She paused a moment to allow that much to be assimilated before she continued.

"Ma'am?" The interruption came from Crewwoman Soderburg.

"Yes, Soderburg?"

"If the ship we're chasing has its own gate generator, how are we supposed to catch it?"

"*Yargus* also has that same system installed, though we have not yet been able to use it successfully." She paused again, allowing a grin to spread across her features for the first time that day. "And now, you all are going to build one for *Dagger*. Lieutenant Reed will oversee the project, with the assistance of Ensigns Hayes, Leslie, Mitchell, and Keone. You will all meet with these officers and Lieutenant Grant at 1300 tomorrow.

"Dismissed."

There were other questions, but they went unvoiced. All recognized the command as the close of discussion that it was. The officers at the table stood and the line of those punished turned sharply to the left and began exiting the room at a stiff march, the Marines moving out of the way to allow their egress. When they were out, the officers followed, Lieutenant Reed glancing at the captain with an expression she had a hard time deciphering. The Marines went last, leaving Captain Johnson alone with her exec.

"You know, James," she said after a moment, "there are far too many ways this could blow up in our faces for my peace of mind."

"Yes, Captain," the XO agreed easily, "but of all the options we could come up with, this seemed to have the best chance of success. And, if I might say, the captain was particularly brilliant in giving the old crew just enough information to generate lots of questions, while also pointing to the new crew as the source where those questions can be answered. It's almost certain that fact alone will get the two groups talking to each other."

"Oh," Fyonna said, startled, "I hadn't thought of that!"

"Of course you didn't, ma'am," Grant countered, knowingly.

CHAPTER 19
1 September
Dagger

Ensign Josiah Mitchell entered the conference room on deck four at a run and slid to a more decorous pace as five sets of eyes turned to assess the newcomer. He saw sympathy on the face of his friend, Jordan Hayes, as his own eyes flicked up to the chrono on the opposite bulkhead. He was only four minutes late, but he could see by the strained countenance of the XO that he was in for it.

"Sorry, sir. I got turned around on deck three and took the wrong corridor. It won't happen again, sir," he gushed without pause as he stood at attention behind the only unoccupied chair at the table.

"At ease, Ensign. Be seated. I'm sure this is not the last time that you will get turned around on *Dagger*. When your watch is completed, why don't you draw a diagram of the corridors and compartments to help you remember your way? You can drop it by my office at 0700 so I can see if you missed anything."

Mitchell swallowed hard and glanced up sharply at the suggestion. He knew it for an order and also knew that he would get no sleep tonight in order to comply.

"Aye-aye, sir. That would be very helpful, sir," he said in the properly respectful tone that allowed no trace of resentment or sarcasm.

He slid into his seat and looked at his friend on the opposite side of the broad conference table. He got a shoulder shrug and a grin in reply.

"Very well, let's get started," Lieutenant Grant said as everyone settled. "I have a few instructions for you before we bring in the enlisted personnel. First of all, ensigns, work assignments have been purposefully mixed to integrate the existing and incoming crews. This is going to cause some friction and it is your responsibility to maintain order," he began and swiveled his gaze to all of the junior officers in emphasis. He lingered slightly longer on Ensign Keone, but only slightly. "Many of you are still very junior and inexperienced, so you will need to be on top of the situation constantly." His eyes again scanned the assembled officers and this time they seemed to linger on Ensign Mitchell. "No excuses will be accepted. You will do your jobs. Is that understood?"

"Yes, sir," they all responded in chorus.

"Very well. Ensign Keone, will you let in the rest of the project staff, please?"

"Yes, sir."

The officers at the table sat very still as the enlisted crew entered and stood crisply at attention against the wall at the back of the room. The crewmen avoided looking at each other.

Lieutenant Grant let them settle for a moment and then began. "The captain has given us an assignment. During our pursuit of *Pathfinder*, she would like us to refit our engines to the *Pathfinder* model to enable us to jump from any location. Lieutenant Reed and I have been going over the schematics and we have broken the job down into four work groups."

Ensign Mitchell looked at the chief engineering officer as the XO continued his explanation. Reed looked wooden and somewhat angry. *What is that all about?* Mitchell wondered. With no way to answer himself, he continued to survey the other officers around the table. Hayes, he knew from their time at the academy and their brief sojourn on *Pathfinder* and then *Vanguard*. They had been friends since they met during their first hellish week of Geopolitical Entities, when Admiral Tudor had put them together to assess the strengths and weaknesses of the old Monarchies and report to the class. *In fact*, Mitchell thought, *I probably couldn't have survived the stress and hardship of the* Vanguard *escape without his support and strength.*

Sitting next to Jordan was a small ensign with sandy blonde hair and bright blue eyes. Her jaw was set as she listened intently to what the XO had to say and she was occasionally marking notes on her e-pad. She seemed familiar and Mitchell thought she had been a year ahead of their group at the academy. The last officer was sitting to his right. While also an ensign, he seemed much older than the others and Mitchell had no recollection of him at the academy. He was neither tall, nor particularly large. His skin-tone indicated that his ancestry probably originated from somewhere in the South Pacific. Currently, half of his face was covered with bandages and the name above his pocket read Keone. He sat with his hands clasped firmly together on the table in front of him and he was looking at Grant intently and ignoring everyone else.

"The first work group will be headed by Ensign Leslie," he said as he turned to the young officer next to Hayes. "Your team will construct a set of six gravity emitters. They are different in design from the Gravitas engine emitters, but fairly similar in function."

Ensign Leslie nodded as he slid a chip folio to her and he continued with his instructions, "You will have Chief Petty Officer Mackey and CPO Giovanni, along with Crewmen Gregory and Adibi."

"Yes, sir."

"Ensign Mitchell," he said as he slid another folio the length of the table to him. "You will be responsible to adapt the current power system to meet the increased demands that will be placed upon it by the new engines. The specifications are listed on page six. You will have Warrant Long and Chief Petty Officer Durrant to assist you," he said with a nod to the two NCOs at either end of the line of people against the wall. "In addition, you will have Specialists Wetzker and Abney from engineering, along with five crewmen, Martin, Bolon, Sydow, Trajanovic, and Corcoran."

Mitchell sat there with a stricken look on his face as he heard the first two names. While he had firsthand experience with Warrant Long and knew that he was occasionally difficult to get along with, he had had no issues at all during their two months aboard *Vanguard*. Steve Long had treated him more like a kid brother than an officer, but there had been no direct conflicts. Durrant, however, had been stirring up trouble ever since they had come aboard. Putting the two on a project where they had to work closely together was like mixing hydrogen and a blow torch. *Why me?* he thought.

"Ensign Hayes," the XO continued in a soft voice as he slid another data folio across to the blank-faced ensign, "you will be responsible to build the needed control systems. The chip has all of the specs. You will have Specialist Paul, and Crewmen Alcaraz, Soderburg, Cuskeran, Trivedi, and Vreeland."

"Yes, sir," he responded with relief.

"Ensign Keone you will be responsible for testing and Quality Control. You will have a crew of four ratings to assist you. They are Van Der Wildt, Fazan, DeGeus, and Loring."

"Yes, sir."

Lieutenant Grant turned his eyes to the group against the back wall and then to the assembled officers. Finally, he spoke in a cold, harsh voice that Mitchell had never heard him use before. "Understand this, people: the captain has been much more lenient than she had to be. In fact, if she had followed the letter of the regs most of you would be headed to the brig for a long stretch. Don't test my patience or hers. The matter is still open until the sentence is completed and as such, can be modified at any time, based on your comportment. A single bad report from your ensign could land you in the brig without recourse. Am I clear?"

Ensign Mitchell scanned the faces of the non-coms and ratings against the wall, especially those who were assigned to his group. Warrant Long stood stiffly with his eyes focused on the far wall in front of him, but his face was red and his jaw was locked. Durrant, on the other hand, had turned slightly to see the XO and the look on his face was just short of murderous. *Lovely*, Mitchell thought. *This is not going to end well.*

"I asked if that was clear?" the XO repeated into the silence with more force in his chilly tone.

"Yes, sir," they all responded automatically.

"Very well. Dismissed."

The officers filed out of the conference room, followed by the non-coms, and the XO was left with just the Engineering Officer, still seated next to him at the conference room table. For a long moment, neither spoke. The Engineer finally broke the silence to ask, "Does she know what she's doing?"

The XO turned his head to respond to the question that should never be asked by a serving officer about their captain. Finally he said, "Tom, you and I have been together for a long time now. You know how inappropriate that question is, but I'll answer it just this once. Captain Johnson is young, but not inexperienced. She has good instincts. After her last tour of this system, she's not exactly as young as she used to be, either. So the short answer is yes, Tom, she knows what she's doing"

"So you think she can handle the pressure?"

Grant stared at Reed for several seconds, incredulous. "Are you serious? Have you read what she's been through for the last three months? Have you heard the scuttlebutt? Yes, without a doubt, pressure is one thing Captain Johnson knows how to handle," the XO said as he stood and left his friend sitting alone in the conference room.

CHAPTER 20
6 September
Dagger

"That makes 110% of safety threshold on the reactor, ma'am. Max accel achieved," Ensign Friedman at the engineering station called to Lieutenant Commander Johnson. She did not see the anxious look in his eyes as he considered adding a warning that Captain Anders would never strain the engines like this.

"Thanks, Engineering. Keep a watch for warning signs on the reactor. We'll back her down below safety levels in 45 minutes, barring any trouble," Johnson called out to the bridge staff. "Steady on the new course, Astro. Comm, patch a message through to Commodore Brighton's operations room and let him know we are running at 110% max acceleration, as he requested," she added.

"Yes, ma'am, message sent," Haralson responded from Comm with only a moment's hesitation.

Johnson went over her display one last time and turned to the Weapons Officer, who was going over something with the staff at Astrogation. "Lieutenant Amaya-Garcia, you have the bridge," she said to the junior grade lieutenant before rising and heading for the hatchway off the bridge.

"Yes, ma'am. I have the bridge," he said, coming to attention and saluting her briefly as she passed by him.

She returned his salute quickly as she moved past. She headed out the hatchway and turned toward her quarters. There were several things she needed to check before the staff meeting Commodore Brighton had called for 1600. It was 1200 right now, and they had just gotten a glimpse of *Yargus* on long range scanners.

They had, as yet, been unable to catch up with her, but considering that *Yargus* was a destroyer, it was only a matter of time before the light cruiser was able to close the distance. It was amazing that *Yargus* had stayed in front of *Dagger* this long. Generally speaking, a destroyer was capable of greater acceleration, but not capable of sustaining a higher top speed, due to limitations on their particle shielding. *Dagger* was two-and-a-half times as massive as *Yargus*, and more than capable of a higher top velocity than any destroyer the WSN had ever built.

The power consumption for particle shielding increased on a low-order exponential curve with the velocity of the ship moving through space. *Dagger* could maintain an appreciably better velocity than any smaller vessel because it had more volume in which to generate power. *Yargus* was obviously performing better than anticipated, as evidenced by their lack of rendezvous. Johnson was sure this was due to the fact that she was of a special design, with higher power outputs than a normal destroyer, and was maintaining a higher speed than any self-respecting destroyer had any business doing. She was familiar with *Pathfinder*'s power budget, but she had never had any reason to push for that much speed. There was a limit to how much velocity you could accumulate before you ran out of star system. Until this one, no ship Johnson had ever served on had any reason to leave the confines of a system.

They needed to catch up with *Yargus* before they got much farther along, however. *Yargus* was engaged in repairs to her new jump engines, and there were a lot of details to be worked out in their plans before they closed on where *Pathfinder* should be. Hence, the increase in acceleration that *Dagger* was performing to try to make up space and reach *Yargus* faster.

She had no doubt that they would. Commodore Brighton had said that 110% would close on them in short order, and she knew that when he said it, he had already done the math. Of course, if they had a working jump engine, catching up would be simplicity itself.

She was pleased at how well, and how hard, her crew was working on the special project. She was thankful that the extra duty seemed to be doing what she had hoped, keeping everyone very busy and thus less likely to cause problems, and forcing many of them to find ways to work together and become more united as a team.

She moved into her study without the practiced ease of familiarity as the hatch slid shut behind her. As captain of a Warner ship, her study still seemed far too large for a single person. In fact, she'd never had a study of her own before. On a light cruiser, and any ship at least as big, there were studies for the captain, XO, bosun, and the heads of engineering, tactical, and security departments; none of which positions she had ever held before. It was another reminder of just how unprepared she was for her current role.

She pushed the unhelpful thought aside and immediately began running the queries to make sure she had all the information in hand for the staff meeting. She had a lot to cover, but she knew Commodore Brighton would not have asked her to provide certain information at the meeting if he hadn't been confident in her ability to get it right. Even though that was the case, she was more than aware that he would know if any of it were incorrect. It absolutely had to be right the first time, since lives hung in the balance, something both Brighton and Johnson herself were keenly aware of.

She settled in for what she knew would take at least the next two hours. That should leave her a little time to freshen up before the meeting, as well.

* * * * *

She met her XO at the lift a little more than two hours later. Lieutenant Grant keyed the destination and the lift surged upward. She was not at all surprised to see Major Chowdhury with the head of ship security, Lieutenant Vitek Gvozdzius speaking quietly in the hallway outside Ops. Knowing the Major, Chowdhury had probably been up here for the last hour. She would want to evaluate everyone coming by before they got inside. The major took nothing at face value, especially not anymore. Johnson was glad to see her, and managed a nod and half smile as she and Grant passed the two Marines on their way into the room. Chowdhury nodded back, but without a smile.

Fyonna remembered that she had wanted to discuss something with her, but just as quickly decided that this was not the time or place for that particular conversation.

Johnson entered the Ops Room to find the large table, usually full of the evidence of the planning and re-planning which took place here, neatly cleaned off. The table was a newer version of the tactical tables in many ships of the WSN. It was egg shaped, with the broader base end of it, on the near end of the room, truncated across in a straight line. Vid screens that could fold up to be visible, or stowed away below, were set at each seat. The tabletop had visual insets in it, allowing for it to function as a display as well as a table. The sections between displays were usually crammed full of printouts and data files. Brighton never had a mess; at least no one would tell him as much. What he had was an amazing amount of information in a limited amount of space. He had apparently cleared everything away to have the table free for the meeting.

The Shipboard Operations Room, its official title, was a medium-sized room, big enough that it could accommodate twenty people. It was constructed in the shape of an elongated octagon, and was built near the communications and tactical core of the ship, with all data feeds duplicated to this room as well as to the main and auxiliary bridges. The three-faced wall at one end was covered with mounted tactical and deep space lidar displays, all of which were currently dark.

The other three sided end of the ops center housed a smaller command room, meant to be used by the ranking operations officer when privacy or solitude was needed without being away from the operations room. Commodore Brighton was in that small command room, at the far end from the hatchway where Johnson and Grant entered. Shortly after those two officers had seated themselves on one of the long sides of the table, Ensign Roberts entered, carrying a set of data chips and was followed closely by the Marines.

Ensign Roberts adjusted the display at the far end of the table, elevating it and facing it back toward the head of the table. Roberts then nodded to the Marines as she keyed in the commands to establish an encrypted laser comm channel with *Yargus'* waiting commander. The vid screen showed the acceptance code traffic from *Yargus* within a matter of seconds, meaning that *Dagger* was continuing to close in on her elusive companion.

Once the transmitted image resolved into the ops room aboard *Yargus* and the resolution began to clear, Roberts went to the command room and let Brighton know that all was ready. A few minutes after Roberts had returned, Commodore Brighton entered the main chamber. The five officers in attendance stood and saluted, and the two a few light seconds ahead aboard *Yargus* stood and saluted on the vid screen as well, a number of seconds delayed in their response.

Commodore Brighton returned the salutes smartly as he made his way to his chair at the head of the conference table. He nodded to all, as a signal to be seated, and took his own seat. All of the arrayed officers did so as well. Ensign Roberts was seated just to the left of the commodore, on the same side of the table as Johnson and Grant. Major Chowdhury and Lieutenant Gvozdzius were seated on his right side. The monitor displayed two people, apparently seated at a desk or table facing the visual pickup.

Commodore Brighton sat forward a little and addressed the group, "I appreciate all of you making the effort to be here and to be prepared for what we need to discuss in this meeting. Commander Johnson, thank you for the continued loan of your Operations Room to my staff for planning. Are you all ready to begin?" he asked of them. He turned to each member of his command and got an acknowledgement before proceeding. He spent a moment looking at each in turn as he scanned them. He liked to start meetings this way; it gave him a brief period to gauge each attendee before he started. He had discovered he could read a lot in a few seconds from people when he had to.

Johnson appeared outwardly eager and poised as ever. Her eyes betrayed a little edginess that was likely due to the enormity of the data he had asked her to analyze for this meeting. Her XO, Lieutenant Grant, was confident, but a little unnerved by the entirety of their assignment on this mission, if Brighton read his engaged expression correctly. That was probably good; it meant he had a better grasp of the situation than Brighton had initially thought.

Lieutenant Commander Julio Ramirez was almost unreadable. It was harder on a vid screen to pick out those nuances of a person, but he looked confident, bordering on cocky. Brighton had to assume, then, that their continuous testing of the jump engines systems had not uncovered any problems. Brighton's mind jumped momentarily to the old joke that those who were able to maintain their confidence and calm when all was falling apart around them, didn't understand the enormity of the problem. Brighton forced the thought from his mind and continued around the table. Ramirez' XO, Lieutenant Jenice Lewis, had been unknown to him prior to this mission. She appeared bright, enthusiastic, and determined, but was perhaps overly energetic. Her face said that she wanted to get on with this so the 'real part' could begin. That kind of attitude might be problematic if she did not have a proper attention to details.

Major Chowdhury, alas, gave nothing away. She had fixed him with that completely attentive yet emotionless face that she oft-times sported and nodded her assent. He shifted to Lieutenant Gvozdzius, who was earnestly watching everyone else, but managed to look back to the commodore and give a slight nod at the moment that Brighton's attention rested on him. He was a good Marine, and according to Sheli, one of the best in the Security Branch that the WSMC had to offer. That was enough for him.

Ensign Jherri Roberts was busy keying up the various data and analysis records he had requested. She was bright and was likely one of the better organized thinkers he had seen come out of the academy in a while. She was still young and inexperienced for her position as his aide, but she was proving herself capable and he had found no fault in her work.

After he had finished getting everyone's attention, he asked Ensign Roberts to provide the status report. She itemized the basics about their positions, and where they were expecting to find *Pathfinder*.

She ended with the statement, "*Pathfinder*, by our best estimates, will be in a position for us to overtake her in 34 days." She nodded to Brighton and then lit up another display at center table, making it visible in 3-D to all of those in the meeting, including those on *Yargus*, as Brighton stood and took control of the meeting.

"The rub, ladies and gentlemen, is that *Pathfinder* will see us coming here," he indicated with a keystroke that pinpointed a position curve, "about 29 days from now." He paused and looked out at all of them. "I asked Captain Johnson to generate some possible scenarios that we will go over later. For now, I would like to hear a readiness report from the Marines on boarding actions and then the same readiness report from Captain Johnson on ship to ship actions with a goal toward disabling *Pathfinder* rather than destroying her. Major Chowdhury, please begin," he asked, reseating himself.

She nodded and inserted her data chip into the slot at her seat on the table. The 3-D display changed to reflect three objects, their two ships and the planet Le Vesconte. She indicated the forces deployed in each ship, the greatest bulk of them being Panther Dragoon on *Dagger*, and then the Platoon left on Le Vesconte to tend the prisoners.

"We are working out some ideas for how to breach and take *Pathfinder* with a minimal loss of life on both sides. Lieutenant Gvozdzius has been working with us, and will continue to do so, providing insight into circumventing certain security features and generating variable plans for the incursion," Chowdhury paused to nod towards Vitek.

"One thing I want to reinforce, though, is that whatever we decide to do during those five days, we cannot discount the efforts that Gunny Aichele will be undertaking on *Pathfinder*, if he is able. He is extremely resourceful. Even if he has managed to only bide his time until then, when he catches wind of the fact that there are WSN ships in pursuit, he will pull out every trick ever written, and write a few new ones himself, if needed. I have no doubt that he will be able to slow them up, maybe even keep them from jumping out of the system during that period, again, if he is still free to act. We cannot be certain one way or another, and so should plan on the pessimistic side," the Marine major finished.

"Major, I hope you are right about Aichele's abilities," Ramirez said. "I am sending some data right now from some of our jump test simulations using our actual test data. It indicates that most likely we will not be close enough to plot the jump that *Pathfinder* might take for the first three of those days. If Aichele, or any of the others, can give us that much, *Yargus* should be able to get a fix on where they are jumping to," he said to the assembled group.

"That is better news than I thought, Captain," Brighton said. "Before we finish here this evening, we will have explored several contingencies and have solid tactics decided upon in advance. I realize that there will be a few things that we will not anticipate, but I want to emphasize my belief and conviction that they had better be few. I want your brains active tonight, all of you. I want to come up with all the scenarios and options that seem likely or marginally likely. Given the time we have available to us before we make contact, we will discuss our response to those unlikely avenues, just in case.

"I want all of you to know and completely understand my position. We will not deviate, we will not back off, and we will not accept any other outcome. We *will* recapture *Pathfinder*. We *will* capture the pirate crew aboard her. And I vow to you all, we *will* bring the traitor, Teach, and those who followed him, to face justice!" The entire room was silent as all nodded in understanding and commitment to his determined cause.

"Now, ladies and gentlemen, let's get to this. I intend for us to cover several things this evening, so we have no time to lose. Captain Johnson, would you be so kind as to share your scenarios with us?" Commodore Brighton said, turning to face his flag captain.

CHAPTER 21
12 September
Dagger

Deserted corridors of the vast warship had their lights dimmed to give its living occupants the illusion of a cyclical night routine. The human mind craved the familiar, and the human body functioned better when such could be found, or simulated. Dutifully, from 2300 each night to 0600 each morning, the ship's lights were set to give off only 60% of their full illumination.

Echoes of work being performed and bits of conversations drifted to Commodore Brighton as he moved, giving an eerie, half-living, half-dead sensation to the scene. The commodore had just left his quarters, and still found himself in the midst of "officer country." Most everyone whose quarters were in this section was asleep, having enough seniority that they were not assigned to take posts in the middle of the ship's night. The immediate area was deathly still, yet those echoes reminded him that the ship was indeed alive and carrying him toward reclaiming the property that was entrusted to his care. The just completed joint planning session with his senior officers made it more difficult for his mind to rest instead of the opposite. He had hoped that the formulation of plans and the setting of specific tasks would have a calming effect on him, but such was not the case.

He wandered aimlessly for some time, with no real course or destination in mind. Something was bothering him, some niggling little detail that his subconscious mind insisted was important enough to deny him sleep, yet not clear enough that his conscious mind could grasp its import.

He left the ladderway when he reached G deck and headed aft, not because he decided to, but because those were his only options; this ladder was too far forward to continue on to H deck, and G deck did not extend any farther forward.

This was not the first time something had been bothering him without him being able to identify the source. The last instance of this nagging feeling, though, had not ended well: two months of cramped quarters, 23 billion kilometers of mostly empty space, little food, little rest, and two of his crewmen dead at the hands of the DaGamans. He hoped this instance would be much less dire.

Problems invariably come up with any command. Brighton's preferred method of relaxing enough to bring out the hidden worries was to review the ship's logs. The methodical examination allowed his conscious mind to fall into a routine where it was busy, but not fully occupied. His subconscious invariably found a way to highlight the problem for him while his concentration was elsewhere.

This inability to drop off to sleep was not normal for him. Slumber had come almost immediately for him, ever since his time at the academy, when fifteen minutes between classes meant ten minutes of sleep. It had been a survival technique that had allowed him to keep up with his studies and remain sane at the same time. Once his body had been trained to waste no time in reaching unconsciousness, the knack had remained with him.

The only time it didn't work, was in instances like this, where his mind resisted sleep until the problem bouncing around inside it could be resolved. Normally, he would review records, but this time he had opted for another solution. It had also been his normal routine to know everything about a ship before taking command of it. This time, he found himself with four ships assigned to him, along with that many more crew and officers as well. And the hectic schedule of an immediate departure from Earth had not allowed him to prepare for his new responsibilities as well as he would have liked.

From a wardroom ahead on the left, G-134 his mental schematic informed him, he could hear some late-night argument in progress. The easily-identified voice of Steve Long hammered the air with blunt force.

"Are you out of your freaking mind, Durrant? What, the laws of physics should rewrite themselves because *you* said so? You can't build a capacitor that will discharge that fast. Even with pure ohmic contacts, you're still…"

The voice faded with distance as he ghosted past and through the companionway, completely unobserved. He shook his head with a reluctant pity for whoever was on the receiving end of the Warrant Officer's caustic comments. Having spent several weeks confined with the acerbic maintenance chief, he knew both his competence and his sharp edges. They had had several confrontations during their short journey. With a slight chuckle, Brighton reflected that Long would probably have had a much smoother career if he had been an officer. He was bright, decisive, and firm in his decisions. He was also combative towards anyone put in authority over him. Brighton let the thoughts recede as he took another ladder upward before reaching the engineering section, which would be cluttered with watch-standers, even at this hour.

Issues rumbled here and there across the surface of his mind as he wandered through E deck, mostly the mundane details of his daily responsibilities. He recalled a scene on the bridge that morning, yesterday morning he realized, checking his chrono to note that it was nearly 0200, that suggested some weapons training was in order. Nothing live-fire, that might give away their approach to *Pathfinder*, but something to shake the rust off.

By 0240, it was clear that he had opted for the wrong method of relaxation. He could still feel the indistinct something his mind wanted him to grasp hovering just out of reach. The last few hours had brought it no closer, and he made a direct course for his quarters.

Spartan furnishings greeted his return. Everything in his living space had been issued with the quarters themselves. Those personal touches everyone adds were not available to him. He hoped to reclaim his effects in the near future, though, in truth, he tended to display very little in the way of mementos: a medal from his academy graduation for the third-highest finish all-time in the Astrogation Trials, a brass statue of WNS *Torshan* which his father had given him to celebrate his first command, and a framed Letter of Commendation for meritorious service during the Humboldt Incident, signed by Admiral of the Navy Franklin Jones. It still amazed and galled him that the Warner Naval Board viewed his actions in that light...after the dust had settled, at least. They certainly hadn't initially.

Still, all these things were a part of himself that he brought to each new command, and without them, he felt like a guest in someone else's room.

He shuffled the thought aside as being unimportant, as well as unhelpful, and sat down before his data terminal. Judging that the important bit was part of something he must have seen in the last week or so, he brought up a list of all the reports he had read during that time. The catalog was extensive.

Reading through the list, there was no way to identify where the information would be found, and so he began at the top of the record and worked his way methodically downward, opening and reading each document closely. The first document was a summary report of the weekly inventory of galley stores. These had been brought to a level sufficient for a one year deployment, far more than the standard two months' supply usually employed by all WNS ships. There was no telling how long this chase might take, and no one wanted to have to give it up because they were low on supplies. Brighton chuckled to see a note from the quartermaster indicating that the supplies were being used at a slightly higher than calculated rate. He himself was contributing to the higher consumption. Perhaps a few more weeks and he would recognize himself in the mirror again.

The second was a routine notice from the Betre gate controllers acknowledging receipt of the comm buoy *Dagger* had sent through the comm gate and giving an expected delivery time to Warner Headquarters on Earth. The time specified had come and gone already, so Admiral Cosina should be up to date on events in the Antoc System, and he would also have a copy of all the additional evidence they had uncovered in the DaGama base.

The third item was a report from Thom Marshall, forwarded to him by the admiral. Marshall had begun an investigation into the death of Captain Vanderjagt and Lieutenant Sepulveda over a year before, and his investigation had spread out to include other related events since. The contents seemed unfamiliar to him as he worked through the known facts item by item. He realized abruptly that he had ignored the message when it arrived because he was too focused on reaching *Pathfinder* to worry about what was transpiring back on Earth.

That wasn't like him, and it was probably his conscience reminding him of duties not performed that had kept sleep at bay. Well, if that was all, it could be speedily rectified. Tunnel-vision seldom brought you the results you wanted, even for the single purpose to which it was directed.

Ignoring the self-evaluation, he focused instead on each point in the report and worked to fit it into what he already knew. Perhaps he could paint the larger view in his mind and use it to guide his plans here in his area of responsibility. Each datum filled in more of the puzzle, yet some indicated that the puzzle itself was larger than he had previously suspected.

The report confirmed a connection between Jhonsruud and the DaGamans that should have been picked up in his clearance checks but had not been. More evidence showed a number of activities that the DaGamans had been involved in. These included ships that were not in any known port for long spans, construction materials that never arrived at their nominal destination, and even what had appeared to be harmless requests for texts and training manuals from Warner and others' military schools.

Marshall suspected that these indicated the DaGama Family had established an extraterrestrial base of operation, and that the purpose, or one purpose, was to start to build an independent military, both ground and space. The report that Brighton himself had brought back to Earth had confirmed much the same conclusion, but he now had information to add to that. Somehow, DaGama also had control of a system which linked to Antoc's primary jump point, and they had constructed at least a small gate generator there. Without those two things, they would not have been able to continue to send resupply pods for the base they had been maintaining in the system.

Other items in the Marshall report indicated links between DaGama and Forrest, and possibly (though not confirmed) similar links from Forrest to other Families. Those links, if they were there, made this a much larger problem, especially if all of those Families were working toward building up a military force outside of Combined Fleet as well as establishing illicit bases that had apparently not been suspected previously.

The fact that at least one other Family was involved was a reasonable assumption from the events Brighton had witnessed in Antoc a few months before. The theft of *Pathfinder* and the attempted theft of *Vanguard* were clearly not run in parallel. If the DaGamans had been after *Vanguard*, then someone else had to be financing the taking of *Pathfinder*. The information that Chowdhury's team had uncovered and passed on to Earth would add a great deal of evidence to Marshall's suspicions, as well as add names to his list of conspiring Families.

Given the content that the commodore was reading now, the Forrest Family was the most likely Family to be responsible for *Pathfinder*'s theft. Their disquiet at the current state of interstellar trade was the most vocal. Their large investment in the production of heavy equipment, mostly for construction and agricultural purposes, made this inevitable. These were items that could be manufactured most inexpensively on Earth with its more refined production infrastructure, but were most needed at the fringes of human expansion. Therefore, Forrest did a lot of shipping, and had its own fleet of cargo ships. What that Family did not own were the gate generators at the various jump points, which they had to pay tolls to use.

Brighton could see clearly the financial reasons for wanting a way around the tariffs. And bottom line, that was the only reason a Family would do anything at all. If it didn't make good business sense, then it just didn't make sense.

As Brighton continued to fit pieces into the expanding puzzle, he began to see a pattern. Each new bit of evidence of the convoluted connections being forged showed that taking a new technology away from a competitor was only the tip of the iceberg.

Ever methodical, Brighton's mind focused on the patterns and, step by step, followed the chain of events being set in motion.

In order for them to gain from the theft of Brighton's ship, they would need to be able to hang on to her, at least long enough to reverse engineer the systems. There's no way they could keep the theft a secret for long, nor any court that would give them leave to keep her. The only other way to keep her would be either to secret her away where no one could find her, or to fight to keep her.

In order to utilize the first method, they would have to have their own system no one knew about, and for the second, they would need to field a naval force of some sort. To improve their chances of success, they would need both.

Theoretically, it should not be possible for them to have gained either. On the initial point, Forrest, DaGama and all the other home-system Families, as those who had not bankrolled the exploration and claiming of extraterrestrial systems were called, had neither exploration ships nor unmonitored access to the jump gates. This should have precluded them from reaching any new system first, and thus filing a legitimate claim. The five original space-faring Families had laid claim to the first five identified jump points, those in the Sol System itself. Each had gone its own way exploring its own new systems, and from them finding new jump links and thus new systems. Whenever the competing Families had found the same system, the Family who had found and surveyed the system first was granted claim to it.

One of these Families had overstepped its bounds from greed and a sense of entitlement. The disagreement had led to war, and as a result, there were now only four space-faring Families: Warner, Sterling, Fermi, and Portales. It should have been impossible for anyone else to reach a new system first.

Yet the fact that there was one system clearly occupied by DaGama showed that it was not impossible. And if it could be done once, it could be repeated.

The other point seemed, on the surface, equally unattainable. The Combined Fleet, another innovation resulting from the Interstellar War of the past century, had strict guidelines laid down to diminish the need for a second such war. Each Family was required by law to match any new ship in its military fleet with a ship of equal size and capability to serve in the Earth Forces Combined Fleet. Since only four Families owned territory outside the home system, only those Families had a military naval fleet. Thus, all the ships of Combined Fleet came from those four sources.

Those four Families were the only ones which trained military officers as well, so all the officers serving in Combined Fleet were from one of these four, proportioned easily enough by serving on the ships of their own Family. The Combined Fleet charter included specifications on how and when an officer was to serve, the main requirement being that an officer could not advance in rank if their CF service time was less than 45% of their total service time. Also, service terms were either eighteen months or two years, defined at the outset, and an officer could not retire nor be reassigned out of CF without the Family he represented replacing him with two officers of equal or greater rank. Brighton was one of only a handful of officers to have seen that provision enacted.

All the Families of Earth provided people to form the crews of the ships, and for many Families' citizens, this was the only way they could find a position off their mother planet.

The theory behind these guidelines, and a reasonably sound one, Brighton thought, was that by creating a mixed group from among all Families, it was less likely that the fleet would ever be used against any Family or group of Families. In addition, working closely with people of other citizenships promoted a more cooperative atmosphere.

It was equally impossible, in theory, for any one Family to again threaten the rest of humanity by building a military force larger than the CF, which received its orders directly from the Ruling Council. If any were to attempt it, the buildup would undoubtedly be discovered before very long, certainly before anyone could obtain a numerical superiority.

Impossible and theoretically impossible are not equivalent statements, however. If Forrest had engineered the theft of *Pathfinder*, then, logically, they must have prepared a way to keep it. The most reasonable preparation they could have made would be for a place to hide the ship, and other ships to defend their prize. Given enough time, and isolation from prying eyes, they could have built themselves a Navy. Their shipping interests would provide them with crews for such ships, though officers would be without the benefit of military experience or training.

Every one of the anomalies in Marshall's report and Chowdhury's summary of her findings supported exactly that scenario. The thought provoked a shiver of dread in him. Forrest was not simply preparing for new ships to avoid tariffs. They were preparing for war.

He tapped a key on the side of his working desk.

"Bridge, Lieutenant Russell."

"Lieutenant, this is Commodore Brighton. Inform the Comm Officer of the Watch to prepare to send a Fleet Priority Message. I will meet her in the communications room in one minute."

"Aye, sir," he acknowledged. Brighton could hear the surprise in his voice. FPM's were not frequently used, and generally, anything important enough to require one was also drastic enough that everyone on the ship already knew about it. The lieutenant at least managed to refrain from asking any questions.

It was a good thing, because the commodore had already left his quarters, this time striding quickly down the corridors with a distinct purpose in mind.

CHAPTER 22
12 September
Dagger

Ensign Jherri Roberts slid her tray to the end of the shelf and scooped it up. The officers' mess was busier than she had ever seen it before, but that wasn't saying much, since this was the first time that she had arrived for a meal during an actual mealtime. Being Commodore Brighton's staff ensign made for a chaotic work schedule, and usually she ate something on the go, if she remembered to eat at all. She quickly spotted an open seat at the far end of the room and smiled to see familiar faces seated at the table.

"Hey, Jos, Jordan," she said as she approached, "mind if I join you?"

"Jherri! Please, sit," Mitchell said, scooting his tray further to one side to accommodate hers. Hayes had two trays at the table, but he had already inhaled the contents of the first, so he stacked the second on top to be sure she had enough room.

She took the offered place and dug into the midday offering with gusto. She realized she had loaded more food onto her tray than she normally would have, but that did not bother her at all. Roberts was still twelve kilos below her ideal weight, and the long hours and missed meals had combined to slow her recovery.

Jordan and Jos looked much nearer their earlier selves than she did, but that didn't surprise her, either. Josiah had a naturally slower metabolism, and he had endured the short rations of the last few months better than most of the rest of *Vanguard*'s complement. And Jordan could eat twice as much as anyone should be able to, and do it twice as quickly to boot.

"So how is it going, working on Captain's staff?" Jordan asked with a grin. When she looked up from her salad, she realized that he was already done eating, while Jos was still working his way through his entrée. She didn't try to correct his use of Brighton's former rank. In fact, she didn't even notice. 'Captain' was the name by which she still thought of him, nine times out of ten.

114

"Well enough," she said, once she swallowed. "He certainly keeps me busy. I wouldn't have gotten away long enough to eat except he asked to speak privately with Captain Johnson over lunch."

Both of the other ensigns looked at each other with raised eyebrows. "What's going on?" Jos finally asked.

"I don't know, and that bothers me," she admitted. "Captain doesn't do a lot of talking about what he's thinking. But something is definitely going on. Did you hear about the FPM?"

"We got an FPM today?" Jordan asked, trying to hold his voice down.

"No, we *sent* an FPM today. The commodore apparently had something quite urgent to say at 0430 this morning."

Jos was puzzled, and it showed on his face and in his voice. "What information was so urgent that we needed a Fleet Priority stuck on it? Did something happen last night that I didn't hear about?"

"Not that I could find out," Roberts replied, "and believe me, I tried. Like I said, I don't know what's going on and I find that disturbing."

Silence held after her statement for almost a minute. Hayes broke it with, "Well, I'm glad I finished eating before you told me that."

"Why?"

"I probably wouldn't have had an appetite after that."

Jos snorted. "The outbreak of another Interstellar War couldn't affect your appetite. Glutton."

"Do I look like a glutton to you?" Jordan asked Roberts, feigning injury.

Jherri cocked her head and considered, but said only, "Hmm."

"Baseless aspersions!" Hayes averred. "I never take more than my share." Mitchell snorted again. "Almost never," Jordan amended.

All three of them laughed, and it felt good to Roberts. There had been too little to laugh about for quite a while.

"So what have the two of you been up to?" Jherri asked, "I'm sorry I haven't spent any time in the wardroom since we left Earth."

"This ship has a wardroom?" Jordan asked.

"Of course it does," Mitchell answered, pulling a folded sheet from his pocket. "Let me show you."

"It was a joke, nitwit."

"You say that *now*."

"Guys," Jherri interrupted, "if *this* is what you do all day, I can find work for you."

"Oh, we've got plenty of work, thank you very much," Jos said. "In fact, we're not due to have off-duty time to ourselves for another four months."

"Why? What did you do?"

"We became ensigns," Jordan said.

"It's not what *we* did, it's what *they* did, and what we have to make sure they do about it."

Roberts made a quick review of the sentence, but found no actual information in it the second time through, either. "What?" she asked.

"You heard about the big fight?" Hayes asked.

"*Everyone* heard about the fight," she confirmed.

"And the Captain's Mast?" Mitchell clarified.

"Yes, of course I did."

"And the punishment?"

"Yes. Everyone got at least a thousand hours of extra duty trying to build a jump engine without a dock or refit. Not likely to work, if you ask me," Roberts stated flatly.

The two young men exchanged a glance. Jordan turned back to her and said, "We have to make sure it does."

"Oh," she said, abashed. Then, "Good luck with that."

"Thanks ever so much," Mitchell gushed.

"Do you remember Captain Black's course on Leadership and Motivation?" Jordan asked.

"No, I took it from Captain Chakrikrishna instead. Why?" Jherri asked.

"I was just remembering that he warned me about this. He said that no matter what punishment you assign to a subordinate, the officer gets the worst of it by having to enforce it."

Jos laughed. "I remember that. He said he still has the same problem with his kids."

Hayes grinned at his longtime friend. "And Jos here definitely drew the short straw for enforcing punishments."

"Don't remind me," Mitchell grumped.

"Why? What's your assignment?" Roberts asked as she polished off the first of the three desserts she had grabbed, a caramel apple crisp.

"Do you know Senior Chief Durrant?" At Roberts' head shake, he continued. "Imagine, if you will, Derrick Mackey's exact opposite in terms of personality. Now put him on the same team as Steve Long and stipulate that no progress could be made in designing a power system until they agreed on each and every point. And add to that the fact that either of them could fold me in half a couple times and stuff me in their back pocket."

Jherri chuckled. "Well, at least the fact that you outrank them should keep you safe."

Mitchell frowned at her. "This *is* Long we're talking about. And Durrant's no better."

"Maybe you should spar with the Marines like Jherri, buddy. They might be able to teach you enough to survive this tour," Jordan offered. Ensign Roberts had started training with Chowdhury as much as she could, even while they were still on *Vanguard*. An odd time to start an exercise regimen, she knew, when you didn't have extra energy for much of anything. Eventually, she recognized it as a means of dealing with the guilt of letting Le

Vesconte die in her arms. Or, rather, it had been recognized and pointed out to her by Major Chowdhury. By that time, though, she had already found many other good points to the workouts, and so she had continued them whenever occasion permitted.

Jos Mitchell gave an involuntary shudder. "No, thank you," he said. "I'll keep out of Chowdhury's way, now that I have a choice in the matter."

"You're scared of her?" Jherri asked.

"Every sane person is scared of her," Jos replied.

"I'm not afraid of her," Jherri commented, and knew that it was true. It hadn't always been true, of course.

"Then you must be insane," Jordan stated. "Q. E. D."

"Very sound logic, Mr. Hayes," Mitchell said in his best imitation Academy professor voice. "For once."

Hayes ignored the jab and spoke instead to Roberts, his curiosity piqued. "Why aren't you afraid of her? I mean, anytime I'm near her I feel like she's counting to herself all the different ways she could kill me before I moved. She looks at me and my lizard brain says, 'Run, run, run!'"

"You may be right," she admitted.

"So you *are* scared of her." Jos smiled.

"No, I mean you may be right that when she looks at you she is actually listing to herself all the ways of incapacitating or killing you." The look she gave them was pure Chowdhury, and they both drew back instinctively. Then she smiled, and they knew they'd been had.

"Anyway, to answer your question, I'm not afraid of her, because I know that she's not a coward."

Again the two youths exchanged looks. Roberts dug into her final dessert, chocolate cake. Finally, Hayes spoke, "All right, I'll bite. What does her not being a coward, which I will readily concede, have to do with her being safe to be around?"

"I never said she was safe," Roberts corrected. "I've never met anyone I considered more dangerous. What I meant was that you can trust her to be in control of herself, because nothing scares *her*.

"Look, she and I talked quite a bit while we were on Gemmill . Not that I'm saying she opened up to me or anything like that, but our conversations did give me a chance to try to puzzle her out. Sometime in her past, she went through something awful. I don't know any details, but it changed her. It burned all the fear out of her. Now, she sees everything differently from everyone else. Her first response to anything new is to analyze it for threats.

"But the way she handles threats is different from most everyone else, too. She doesn't move against a threat until she knows she has to, until she is sure that it is necessary."

"I'm not sure I buy that," Mitchell said. "Okay, I can see that having supreme confidence in your own abilities would allow you to give people more leeway before you had to act. And I can see Chowdhury having that kind of confidence. But how do you go from that to your conclusion that Chowdhury can be trusted not to break me in half on general principles because she's too brave?"

"Have you ever met a bully?" Jherri asked.

"Durrant," both men said in unison.

"Okay. Have you ever heard it said that all bullies are cowards at heart?"

"Sure," Jordan said. "You hear that all the time."

"So why would someone who's afraid of being beaten go out and pick fights?"

"I don't know," Jos ventured, "I don't think they're looking for a fight so much as looking to intimidate you into backing down."

"Exactly!" Roberts said. "They don't want to fight, they want you to be too scared to fight, because they're afraid of losing."

"Hold on," Jordan piped up. "If you're trying to demonstrate the difference between Chowdhury and a bully, why is it that I am totally intimidated by her?"

"Because you're not insane like her," Jos quipped. "Sheesh, we've already covered that. Try to keep up, fruitcake."

"Funny boy," Jherri said with mock gravity. "Come here and let me show you a move I learned from a Marine PFC the other day."

"Truce," Jos cried and raised both hands to ward off an imaginary blow.

"Back to your question, Jordan. I asked the same one myself, so I've done a little thinking on the subject. I think it's true that Chowdhury deliberately tries to intimidate people, and that had me confused, because she's not a bully and she's not a coward. I finally decided that the difference was in the reason she was intimidating.

"For a bully, the goal is always to gain some advantage; your lunch money, your silence, your help, something they want that you wouldn't give up voluntarily. With Chowdhury, the intimidation is for several reasons. First, I think she wants people to avoid fighting with her because she'd rather not hurt anyone if she can avoid it. Second, I think it makes her life easier if people jump when she says frog. And third, I think she prefers being isolated and insulated from others.

"Whatever she's been through, she doesn't let people in anymore," Roberts concluded.

"I disagree," Jordan said.

"Why?"

"Well, back on *Pathfinder*, the only person I would have thought she was at all close to was Captain. But on *Vanguard*, she got close to quite a few of us, you among them. Didn't you just say that the two of you had several talks on Gemmill?"

"A point," she allowed thoughtfully. "Let me amend that to say that she doesn't let people in easily."

"Especially not ensigns," Jos added.

"Amen, brother," Jordan said. "At least not most ensigns," he corrected, and nodded Jherri's way.

"I think you'd be surprised at her opinion of you two, as well, though she'll never tell you herself. I think you're closer to family in her book now than the green officers you were."

Jos scoffed. "Green officer would be several hundred steps higher than what her opinion of us was." He attempted an imitation of her "Stupid Ensign" look, but he couldn't really pull it off.

Jordan could tell what he was trying, but still asked, "Do you need to go to the head? You don't look so good."

"Funny boy," he shot right back. "Why don't you let Jherri show you a move a Marine PFC taught her recently?"

All three of them laughed at that. It was Hayes who eventually brought them back to a serious subject.

"I think I would have to agree with Chowdhury."

"How so?" Roberts asked.

"Well, we were rivals at the Academy, and friends on *Pathfinder*, but I think of you and Jos like my sister and brother now. The same goes for everyone on *Vanguard* with us; they're family now."

"Amen, brother. Except for Delacoeur; I'll never trust that weasel," Mitchell repeated seriously.

"Amen," Jherri said, and she suddenly looked close to tears. Jordan decided it was time to move to a lighter topic.

"Speaking of family, did you hear about the wedding?"

"Yes! I got an invitation in the last mail call before we jumped out of Betre," Jherri said happily. "Leon and Elle sure didn't waste any time getting engaged!"

"No, they didn't," Jos said. "But they were kind enough to set the date for June, to give us every chance to be done here so we can attend. He asked Jordan and me to be groomsmen, too."

"No kidding? That's great. I'm very happy for them."

"Do you think June is going to be enough time to recapture *Pathfinder*?" Hayes asked the other two.

"Depends," Roberts answered. "Captain expects that there's still a fairly good chance that they haven't left this system yet. If the commodore has calculated the odds of something, I'm not going to be the one to tell him different. Still, I have my doubts. I know the ship was locked down before we left her, but that was almost twelve weeks ago. You'd have to think that they could make repairs a lot quicker than that, even with what we know has been sabotaged already."

119

"Maybe Captain is counting on Samuels and Aichele and Burton to keep them here somehow," Mitchell said.

"That's really all we have to pin our hopes on, isn't it?" Jordan asked rhetorically. "Normally, I'd say the chances were somewhere between slim and none. I'm not even sure they are all on our side. Captain sent Samuels off *Vanguard* but she looked broken. I hurt just thinking of her on that ship with those pirates."

"I agree. Chowdhury told me she sent Aichele back to guard Burton and that he was just as capable of doing that as she was herself," Roberts answered quietly. "If she trusts him to keep *Pathfinder* from jumping, I'm not going to argue with her. And besides, Captain gave his word that he would find them and get *Pathfinder* back, no matter how long it took." Roberts said.

"Does any of this ever remind you of Moby Dick?" Mitchell asked.

"Moby who?" Hayes wanted to know.

"You mean, you think Captain is acting like Ahab and *Pathfinder* is his white whale?" Jherri asked, ignoring Hayes' attempt at humor.

"I hate to say it, and you know I look up to Captain, but he does have an obsessive side to him. Is it possible that he might drive us into more than we can handle chasing after his lost ship?" Mitchell did indeed look unhappy to say it.

"I wouldn't worry about it, Jos," she answered. "For one thing, the current situation is well within our ability to handle. If they do jump, they're probably still going to be in Warner space, where we will still have the upper hand. For another, Brighton is not really the obsessive sort. He is focused, and he is strict, but Captain Ahab he is not. Besides which, neither Major Chowdhury nor Captain Johnson would allow him to act like that."

Mitchell grabbed his tray and rose. The others joined him, Roberts with something approaching panic when she noted the time. "I think you're right, Jherri," he decided finally.

"No, seriously, guys," Jordan asked on his way out the hatch, "who's Moby Dick?"

CHAPTER 23

22 September
Yargus, Dagger

WNS *Yargus* was an old ship. But as old as she was, her utilitarian bridge gleamed with newly-minted components. The destroyer's hull had been commissioned back in 2725 as WNS *Reigna,* and she had seen some hard use in her sixty years of service, but the rapid refit and re-christening she had undergone had upgraded practically every system to the latest available equipment.

Her engines, too, took advantage of both top of the line equipment and practical improvements gleaned from Warner's century-long position at the forefront of innovation. The problem being that life on the cutting edge meant that risk and failure were found as often as not. Captain Julio Ramirez was very much aware of his ship's capabilities and limitations, having most recently served as her Chief Engineer during her refit and workup in an out of the way corner of Warner space. He had risen in rank and responsibility primarily through the engineering track, and this was the finest ship he'd ever served on, bar none, even if he and Lieutenant Burkhalter, *Yargus'* current Chief Engineer, were still trying to debug her jump engines.

His officers were of equal quality; each of them was hand-picked for this assignment by Admiral Cosina, and that officer was not the sort to leave any detail at a less than optimal state. *Yargus* had received an influx of junior officers just before she had left the isolated dockyard for her shakedown cruise. The five top graduating ensigns from the Naval Academy's current class had reported aboard, so green they squeaked, but quickly acclimating and contributing. One of these, Ensign Jeffery Lyle, sat at the helm station, and another, Ensign Jessica Ragnarsdottir, was operating the communications console. The three other ensigns, Craig Judd, Tomiko Kikuchi, and Jared Lee, all held similar positions during the other watches.

His executive officer, Lieutenant Lewis, was currently hunched over the comm board with Ragnarsdottir, explaining the ins and outs of duty at that station to her. Lewis was not due on the bridge for more than three hours yet, when she would take command of second watch. She was a very hands-on individual, though, as he had learned, and overseeing the training regimen for the ensigns had meant that she invested a lot of her personal time in one-on-one instruction. It certainly had been paying dividends; the intelligent and able young men and women soaked up the information and routines more quickly than Ramirez had ever seen before.

The time Jenice Lewis had put into the project did not show any signs of diminishing her performance with her other duties, either. Besides being a personal mentor to each of the new officers, she had a full load of responsibilities to the rest of the ship, in organizing the crew's schedules and assignments, handling the day-to-day issues that continually arise on any ship, overseeing the inventory, managing the various department heads; nearly an inexhaustible list. Yet she attacked each item with her typical energy and enthusiasm and Captain Ramirez was left with a smoothly operating command, and very few troubles that ever came his way.

At least, so far.

Ramirez had a lot in common with his command: he had just been upgraded to the rank of Lieutenant Commander, and he had a reputation as a smooth operator. His shorter than average yet muscular physique was as much a draw to the opposite sex as was his darker complexion. Of course, naval regulations prohibited any sort of social connection between an officer and those under his command. Previously, he had been able to arrange shipboard romances with officers in other departments. Now that he had risen to command his own ship, those days were past.

A fact he regretted somewhat when he took a second look at Lewis. Not enough to give back the promotion, of course. This was one of the drawbacks of the job, but you learned to take the bad with the good in life. No man could have it all, and trying to do so would only lead to disaster, he reminded himself philosophically.

Some who only knew Ramirez peripherally might have thought him obsessed with just that goal. It was clear that the young man was ambitious. He worked long hours and studied everything he thought would increase his chances of advancing himself. Such a vigorous work ethic had allowed him to make few friends and several enemies. His relationships tended to be short-lived as well, never lasting longer than a single deployment.

Those who saw only the surface would have been mistaken, though. Ramirez had been orphaned at the age of four, early enough that he had only vague and ephemeral memories of the family he had lost. Both of his parents and his sister had died in the flooding of Hurricane Michael in '63. That Ramirez himself had survived was a miraculous confluence of luck and the tenacity of his older brother Pablo, who had been nine at the time. Pablo had somehow managed to hold onto him and grab a tree branch to keep from being swept down what had been their home street the day before, but had transformed into a raging maelstrom of churning, swirling mud and water. Pulling both of them out of the torrent and into the tree, they had remained isolated there for nineteen hours until help could find them.

A week later, it had been concluded that no member of Julio and Pablo Ramirez' immediate or extended family had survived the devastation of the countryside. Though not Warner citizens themselves, they had been taken in, along with thousands of other suddenly homeless people, by the generosity of the nearby Warner Enclave in Mérida, which had survived the violent storm with hardly any damage at all.

Within the month, all but a handful of those refugees had returned to where their homes had been and started rebuilding, putting their lives back together as best as they were able. The two Ramirez boys were part of that handful. They had nowhere else to go.

The generosity of strangers again picked up after the whirlwind of fate. Matthieu and Theresa Lyon took them under their roof, and eventually adopted them both. Matthieu had been the top operations engineer when Julio had changed his last name to Lyon. He was promoted to be Mérida Enclave's manager the summer before Julio's fifth grade year, and promoted again to the Warner Board of Directors just after the youngest Lyon had started high school. This promotion also meant relocating, at least for Julio. Pablo had already begun taking university courses in business, so he had remained behind.

Julio went with his family to live at Warner Headquarters in geostationary orbit 35,785 km above the Quito Enclave, and his love of space was born. This was the first great event that decided him on the course his life should take. The second was to come during the next summer.

Pablo returned home for the long annual break from school, but he did not seem his normal self to his brother. When Julio pressed him for the reason for his funk, he confided that he had learned that the high grades he had received were not necessarily earned, that some instructors were inflating his marks simply because of his name.

Julio had thought long and hard about what that meant, and what he could do about it. After talking it over with his father, and gaining his consent, he had enrolled at the Warner Naval Academy using his former name. He never let anyone know that he had political connections, and he knew his father had not used them in his behalf. Knowing that he had achieved his own command solely by his own efforts made the success all the sweeter.

And the sweet taste of success was enhanced further by the trust Admiral Cosina had shown in him when he assigned him to a mission as important as this one.

A Cheshire cat grin that he couldn't suppress emerged as he took a slow look around the bridge one more time. All was as it should be, everyone performing their function without the need for his input.

"XO, take the bridge, if you would," Ramirez said, standing from the center seat. "I'm going to head down to engineering for an update, and to see if I can pitch in. It wouldn't do for *Dagger* to get her jump drive installed and tested before we get ours working, after all."

Lewis straightened from the comm board and turned. "Aye, sir. I have the br—"

"IFF ping," interrupted Ragnarsdottir in a firm voice.

Ramirez stopped at once. "Where away?"

"Source is dead ahead of us, Captain," Lewis filled in before he could arrive to check for that same information. He returned instead to his former spot and sat in the command chair.

"Return ping," he ordered. "Get me an ID, just to be sure, but that has to be *Pathfinder*. Sound general quarters, XO. Comm, signal to *Dagger* to report contact."

"Aye, sir," the ensign at communications responded, much more calmly than Ramirez would have expected. Everyone on the bridge tensed, waiting out the long seconds for the light-speed challenge to be received and answered. The alert signal sounded throughout the ship, calling everyone to their primary duty stations in anticipation of battle.

"Comm signal inbound," Ragnarsdottir reported, rather than the IFF return Ramirez had been expecting. Ah, *Dagger* must have been pinged as well in advance of his contact report.

"New orders, ensign?"

"No, sir," the young blonde responded in a puzzled voice. "It looks like a standard encryption, but the computer hasn't been able to decrypt it. I don't think it's ours, sir." After a few seconds of silence, "IFF reports comm buoy from WNS *Pathfinder*, sir."

Ramirez hesitated over his next orders, because this did not make sense to him.

"Astrogation, give me a distance based on IFF return time."

"Aye, sir. Distance to comm source is one-three-four million km, Captain," Lieutenant Hammerberg announced, without looking up from the astrogation board. He already had the answer before the captain had asked, Julio noted. It was a joy to work with good officers. "Doppler shift indicates we're closing fast, sir," the AO continued.

"How fast, Karl?"

The lieutenant seemed to be calculating still, but said firmly, "Buoy's velocity is steady at two-six-five thousand kps."

"Comm, report our actions to the commodore and ask for or—"

"Orders from Fleet CO, sir," she interrupted again. "We are to recover the buoy and bring it aboard. Staff meeting at 2100. "

"Acknowledge the orders, Ensign." It was a joy to work with good officers.

* * * * *

Commodore Brighton certainly knew what he was talking about when it came to both the timing and the navigation of his orders. Recovering the buoy meant slowing enough that the tractors could get and keep a grip on it. The heavy deceleration allowed *Dagger* to make up the last of the distance separating the two ships. Rendezvous was accomplished at 2042, after *Dagger* had also slowed to match speeds with her junior companion. That had allowed just enough time for Ramirez to board a launch and transit to the first face-to-face conference since their departure briefing. Lewis had remained in command on *Yargus*; regulations demanded that the captain and executive officer could not leave the ship at the same time; but she would join the meeting virtually, and without the bothersome comm lag this time.

Gathered in the operations room when Ramirez arrived at 2102 were most of the same officers that had attended the last such assembly. Brighton was already seated at the far end of the plotting table, and it was clear that some discussion had already begun. He hoped his tardiness had not caused them to start without him. More, he hoped that it had not annoyed the stern-faced commodore who motioned him to a seat.

"Major Chowdhury, if you'd please repeat your last statement for Commander Ramirez, I think we can now begin," Brighton suggested, and the discussion abruptly stopped.

"Yes, sir." She turned to gaze levelly at Ramirez, and he could see that she, at least, was annoyed that he was late and that she would have to repeat herself. After looking into her cool eyes, he decided he should go to great lengths not to annoy her again.

"Commodore Brighton asked me to have a look at the incoming communication as soon as it was detected. Due to my…previous experience, he felt I might be able to decipher the message using the DaGaman codes we acquired recently, or one of the other Families' active codes I have access to. When I ran the signal through my personal checker, it found a match at once. The encryption key used to secure the signal originates with the Warner Marine Corps. The computer is still processing the entire message, and should finish soon, but I can tell you that the message came from Staff Sergeant Jill

Burton and was sent without the knowledge of the pirates who took *Pathfinder*."

"Why not use a standard naval code?" Ramirez asked. "It would have saved us some time in finding out what she had to say."

The look Chowdhury gave him made him want to retract his words, especially after his recent vow not to get on her bad side, but she explained calmly, "Besides the obvious fact that she could only have obtained such a code from a naval officer, who, as far as she would have known, was a traitor, it was prudent on her part not to risk it. She would not have been sure who would follow after, and if one ship can be taken, another might. Also, there are only two people on that ship who could have read the message if it were discovered before it could be launched, and both of them are known to be loyal to Warner. Her faith in the Warner Navy is understandably shaken, and the buoy was directed to Corps Headquarters instead. Her first loyalty is there, and rightly so." Other than Gvozdzius, she was the only Marine in the room, but she glared at each one sitting there, as if daring them to disagree.

Wisely, no one did. For Ramirez' part, he'd wanted to sink into his chair at her slight emphasis of the word 'obvious.'

A chirp at her station broke the silence. "Decryption is finished, sir," she directed to Brighton.

"Begin playback, then."

Chowdhury entered a few keystrokes and then a clear alto voice was heard from the overhead speakers.

"Staff Sergeant Jill Burton, reporting status for WNS *Pathfinder* as of 2 September, 2787.

"On 27 June, this year, *Pathfinder* entered the Antoc System employing a jump gate of its own creation. On 28 June, individuals serving on *Pathfinder* but disloyal to Warner, successfully took control of the ship. Injured in the fighting were myself; Specialist 3 Glenn Morales, disloyal, deceased; Crewman 3c Carl Brandon, disloyal, deceased; Major Sheli Chowdhury, loyal, minor head wound; and Lieutenant Neil Lamont, disloyal, missing and presumed dead.

"Nineteen of the ship's company were removed on *Vanguard* and stranded on planet A3 of the Antoc System. These are: Captain William Brighton, Major Sheli Chowdhury, Lieutenant Fyonna Johnson, Lieutenant Leonard Ward, Ensign Jherri Roberts, Ensign Josiah Mitchell, Ensign Jordan Hayes, Chief Warrant Officer 2 Steve Long, Warrant Officer Elle Williams, Warrant Officer Timothy O'Neill, Master Chief Petty Officer Derrick Mackey, Chief Petty Officer Clémence Quèneau, Petty Officer Drew Le Vesconte, Specialist 1 Jens Fujinami, Specialist 2 Claire Paul, Crewwoman 1c Kara George, Crewman 1c Kieran Delacoeur, Crewman 2c Ricardo Smith, and Crewman 3c Roberto Alcaraz. Every effort should be made to expedite relief to A3. The above named were given neither extra rations nor shelter before being marooned.

"At this time, nine separate acts of sabotage have been noted, but the identity of the saboteur is known for only five acts. Sgt. Aichele disabled the long range communications equipment within the first week after losing the ship. He is also responsible for disabling the jump engines and main engines. The next events were minor damage to the field control station in Engineering, severe damage to the engineering control runs between the bridge and the after half of the ship, the weapons control station on the bridge was disabled, and the astrogation data has been tampered with and is no longer reliable. Other actions have been directed at damaging or reversing the repair work which is being done for the command and control systems. All of these were perpetrated by a person or persons unknown to either Sgt. Aichele or myself."

"Samuels," Brighton interjected quietly.

"Two weeks ago, I had recovered sufficiently to take action against the traitors in control. I have been aided in this by Doctor Meghan Johnson, who has helped cover my absences from the medbay. Over the course of two days, I was able to gather all vacuum suits on the ship and transport most to the exterior portion of hold seven. Shortly thereafter, my absence from the infirmary was discovered and I was forced to hide out in hold seven."

The faces of all those around the table were intent, rapt by the images the Marine's words conveyed. Brighton stared at nothing, but Ensign Roberts' attention was focused on a hand pad where she was taking notes rapidly as the message played.

"Sometime during the evening of 16 August, Ed Teach, who had assumed command, died under unusual circumstances, leaving only three officers on the ship. Katherine Leung assumed command on 17 August. On that date, she also promoted Monica Samuels to lieutenant and named her the executive officer. Stuart Omundsen was put in charge of the Engineering Department."

Brighton's face went pale at the news of Teach's death. He continued to listen, but his eyes held a hollowness that had not been there before. Roberts looked up sharply when Samuels' promotion was noted, but just as quickly went back to her notes.

"Aichele has been under suspicion since his first act of sabotage, and has not been able to arrange another for several weeks. He has managed to discover the arranged meeting place, where Teach was, and now Leung is, planning to turn over the ship and receive payment. The Family funding the theft is Forrest. The location is somewhere in the Worth System."

Several startled glances were exchanged around the table at this, but Brighton only nodded.

"Captain Brighton arranged for the disabling of nearly all ship's systems before being put off *Pathfinder*. The sabotage has slowed repairs considerably.

However, Engineering is now completely functional, at least on local control. Sensors are partially operational. Weapons control is restored. Inbound communications can be detected but *Pathfinder* has no ability to transmit. Aichele and I have been forced to operate independently, and both of us found ways to disable that system. Repairs will be impossible to effect without pressure suits, of which I have deprived the ship. The jump engines will likely be repaired within the coming week, but the lack of dependable astrogation data should keep Leung from attempting to jump anytime soon.

"Gunny Aichele and I have discussed our situation at length, and we feel that we can certainly keep *Pathfinder* from jumping clear for another two weeks. We are equally certain that we cannot delay things more than eight weeks. It will be four weeks before this message arrives at its destination, and it is possible that it will be too late already to keep the new technology out of the hands of the Forresters. We will do everything we can to keep that from happening. A Sigma Option is being arranged, though we cannot guarantee that it can be prepared in time to implement before we jump.

"Semper Fidelis. Burton clear."

No one spoke for several long moments.

"Well," the commodore said gravely, "at least we know that they are still in the system, or were two weeks ago." That said, Brighton leaned back and began tapping his jaw with the index finger of his right hand. He continued that action as discussion once again began at the table.

"It's fortunate that the Marines were able to keep *Pathfinder* here in the Antoc System," Ramirez said, nodding in the vague direction of Chowdhury and her second. She did not comment, nor did she nod in agreement. From her point of view, fortune had nothing to do with it, but she was not about to try to explain it to him.

Johnson did comment. "It certainly makes our success more likely. Had they managed to jump without Burton telling us where they were headed, it would have been nearly impossible to pick up their trail again."

"The information we have now is of enormous value in planning the retaking of the ship, as well," Lieutenant Griggs added. "The commodore's staff and I have been preparing for more than forty potential scenarios. Now we can focus and refine the dozen or so that may apply." He turned to Brighton to ask a question, but hesitated at seeing him absorbed in his own thoughts. He looked next at Major Chowdhury, but did not like what he saw and continued on until he reached Ensign Roberts, the only one in the room he outranked. "So with Teach dead and Aichele, Burton, and Doctor Johnson working against Leung, how many effectives will we need to account for when we board?"

"Assuming Samuels is on our side as well, which I will take the commodore's word for," she said firmly, "there should be two officers and twenty-five crew left to deal with."

"And we have better than forty Marines," he said approvingly.

"Sixty-six," Chowdhury corrected, "if I include the security units from both ships; forty-eight without them."

"And what about the window Burton gave us for when they might be able to jump? Will we be able to reach *Pathfinder* before that?" Griggs asked. Johnson thought he should have already known.

"Twenty-one days to intercept still," *Dagger*'s captain replied. "If they don't alter course or speed up, since they now have their engines back online."

Lewis chimed in, "No reason to think that they would have tried to evade. They would not expect the theft to have been discovered yet, so they won't be expecting pursuit. Plus, the comm buoy was dead on a line from where we expect to find *Pathfinder* and the comm gate at JP1."

"So our basic strategy should stay the same," Griggs opined.

"Absolutely not," Brighton pronounced. "Our current approach was the only logical one to adopt when *Pathfinder* herself was the only lead we had to follow. Now that we are aware of her destination, we have two goals. Continuing pursuit with both ships is favorable for us in only a handful of potential scenarios. If the situation suddenly changes on us, we will have no options to turn to. That is not acceptable. In order to cover more eventualities, this is what we are going to do.

"At 0400, *Dagger* is going to reverse course and return to Antoc," he stated, looking directly at Lieutenant Commander Johnson as he spoke. "In Antoc, she will recover the Marines and prisoners on Le Vesconte and bring *Avram* with her back through JP1. Expedite work on the gate generator as much as you can. If you manage to get the jump engines on *Dagger* working, don't wait for *Avram* or *Foundation*, jump straight to Worth and find *Pathfinder*, if she has jumped, or the Forrest ship waiting to meet her if not.

"If you do need to take the long way around, split *Avram* out at Gateway and send her to Earth with a copy of all our findings, along with the prisoners, a guard, and Lieutenant Griggs." He turned to the named officer. "Lieutenant, I'll want you to brief Admiral Cosina personally on this information."

"Aye, sir," he acknowledged in a voice that was at once proud and disappointed.

"And *Yargus*, sir," Ramirez queried.

"*Yargus*, Captain, is about to become my flagship. My staff will need to remain here, except for Ms. Roberts. We won't have room, because I intend to stuff as many Marines as will fit into her, and we are going to catch *Pathfinder* before she can jump."

CHAPTER 24

3 October

Earth

Admiral Conrad Cosina paced back and forth across his office in front of the floor to ceiling windows that looked out over the Quito Complex. His office was on the forty-first floor of the Navy Building, two below the top floor, which was reserved for the CNO. It was a fitting place for the Deputy Director of R&D to have his office when he was on Earth.

As his job had overlap between the Fleet and the Family at large, he also had an office in the Warner HQ in orbit, some thirty-five million meters above what he considered his "working office." He much preferred the view from these windows to the stark clarity of space. The patterns of lights far below, which were normally so soothing, failed even to register in his consciousness on this night. He turned and glanced back at the data chip occupying the center of his desk. He had read the contents of that chip over and over and the conclusions which it contained. He didn't want to believe that such a thing was possible. The chip contained the sum total of a Fleet Priority Message, one of only three the admiral had ever dealt with directly during his tenure in the navy, a report from Commodore William Brighton which included his conclusions and suspicions based on evidence they had been able to uncover in the Antoc system.

Reaching the end of the stretch of windows, Cosina turned and began another circuit. He was a tall man, nearly one hundred-ninety centimeters in height, but his bear-like stature disguised his height in an overall presence that was quite intimidating. At ninety-three years of age, he had seen most of the wonders of the world and had been instrumental in discovering the wonders of many new worlds, spending the majority of his early career in the Navy's exploration division. For the last decade and a half, he had been directly involved in the projects of the Warner Family which would help the Warner Space Navy and the Family to grow, progress, and, most importantly, to

130

maintain the lead over all the space-faring Families. Because of his efforts, Warner controlled more planets and jump points than any other Family. Due to that fact, according to the regulations established in the Treaty of Dallas, they had the largest fleet in known space. He turned and looked at the chip again. If Brighton's suspicions were correct, they were about to have a serious threat to their status as the most powerful of the space-faring Families. Indeed, they were about to have the most serious physical threat to their property and holdings throughout the settled galaxy since the Vector Rebellion of the previous century.

They needed to get more information, and in a hurry. So far, they had never gotten a hint of any of these suspected activities except those which had been directed at the prototype vessel, *Pathfinder;* but it would seem those incidents had only been the tip of the iceberg. Even without direct evidence, Brighton's inductive reasoning was quite sound. But inductive reasoning is not enough to convict in a court of law; for that, they needed evidence.

"Colonel Valencia is here for you, sir," called the intercom from his desk. He moved to the door and opened it himself.

"Please come in, Colonel," he called and then held the door until the diminutive officer had entered the office.

The two officers were complete opposites. Cosina with his light brown hair graying at the sides, gray goatee and bulky, ponderous steps and Valencia's jet black hair pulled back into a short ponytail at the nape of his neck, sharp beak-like nose and quick furtive motions. They were the bear and the raven. For all their dissimilarities, they shared one common trait, they were each very good at their jobs. Colonel Stefan Valencia had risen through the ranks of the Warner Marines at a steady pace through sheer unending competence, coupled with regular flashes of brilliance, to become head of the intelligence gathering branch of Covert Operations. Covert Ops was under the direction of the Office of Intelligence and Research, which made Cosina his boss.

"Please sit down. Did you receive the folio that I sent over?" Cosina asked as he motioned Valencia into an armchair facing the desk.

"I did. I only had a few moments to scan it before your call, however, so my analysis is very superficial."

"What were your initial thoughts?" he asked as he made his way into his own chair behind the bulk of the massive cherry wood desk.

"First of all, I saw the data on DaGama but I wasn't sure how Forrest tied into that."

"We aren't sure they do. That is one of the things I want you to verify. At this point, it is just circumstantial association."

"Sir, with something this explosive, we need corroboration from multiple sources."

"Absolutely, that will be our first priority and your first assignment. What else?"

"If this is true, and we have to assume for now that it is, we need to reinforce throughout the frontier. We need to sweep each system extensively and in force."

"I'm not sure that we have the forces to pull that off, but I don't disagree. Anything else?"

"Yes, sir," he said and then paused as if his next statement was personally distasteful. "In order for this level of action to be hidden, our security must be compromised at some level."

"I'm not happy with the ramifications of that statement," Cosina leaned across his desk in a visual representation of his displeasure. "Marine Security is not easy to fool, nor to circumvent."

"No, sir, they are not. I am not any happier with this assessment than you are," Valencia said without the smallest sign of intimidation. "I simply don't believe that it would be possible to hide these events without inside help."

Cosina leaned back as far as his chair would permit and turned to look out the window again. He sat for several moments without a word. Valencia, who was used to these moments of introspection and evaluation, waited patiently. Though he, personally, was a man of quick action and nervous energy, he had long ago realized which method worked best for his boss.

"Okay," Cosina said finally. "As I stated earlier, the first priority needs to be to verify this information. He stood and paced back and forth in front of the large window again, but his eyes were on the floor instead of the wonders outside. As he turned at the end of his circuit and began to come back toward the desk and Valencia, he looked up and said, "I hate to do this, but I think we need to keep this intelligence gathering operation completely black. No contact with Warner Corporate Security at any level." He waited for Valencia's nod before continuing. "At some point we will need to check out our own security people, but the information verification has the higher priority. Keep your group as small as you possibly can, but I need definite confirmation within two weeks."

"That's awfully quick. It'll be tough to keep it completely quiet," he said, looking to the admiral for confirmation of his understanding.

"Do the best you can. Speed is the priority, but you need to be very careful. We know of several deaths that can be connected, at least tangentially, to keeping this quiet."

"Yes, sir."

"Any questions?"

"No, sir."

"Very well, dismissed."

Valencia stood and moved to the door without a further comment, his head moving up and down as if he were arguing with himself over the best way to accomplish his orders. Cosina watched him until the closed door hid him from view and wondered what he had just set into motion.

If he were the sort to believe in premonitions, he would have thought he had just started an avalanche that was about to bury the Warner Family - as well as many of the other Families on the Ruling Council.

CHAPTER 25
2-6 October
Yargus

Lieutenant Commander Julio Ramirez, Captain of WNS *Yargus*, unlocked the bridge hatch and entered at 0730, a full twenty minutes before he was due to take over first watch. Not surprisingly, he was not the only officer from first watch to have reported for duty early. Ensign Judd's bulky frame was seated at Helm and Ensign Lyle had control of the Engineering station with an infectious smile on his face. They probably had not been there long, since the ensign and warrant they were replacing were still present, conveying the status information for their station to the new arrivals.

"Captain on the bridge," Lieutenant Shuyler reported while rising from the center seat, and coming to attention, followed immediately by those not seated at one of the bridge stations. It was a longstanding tradition in the navy, but Julio was new enough in his posting that it occasionally caught him unprepared to respond. He braced himself to attention, and nearly turned in place to face the captain entering behind him, but managed to stop himself in time.

"As you were," Ramirez said. "Lieutenant, I have the bridge."

Shuyler moved to one side to allow his superior to take the command station. "Captain has the conn," he intoned.

"Status report, Mr. Shuyler," Ramirez directed. Ramirez seated himself and adjusted the angle of the command repeater screens to face him at the lower altitude he needed.

"On course and traveling at maximum speed. No new orders or communications logged this watch."

"Estimated time to scan range?"

"No change, Captain. If *Pathfinder* has not altered course from what Sgt. Burton reported, we should pick her up at 0812," the lieutenant reported.

"Very well, Lieutenant, dismissed."

"Aye, sir."

134

Shuyler stepped away toward the Astrogation station, rather than toward the hatch. He leaned over to check an item or two, then moved over to Helm and did the same. This proved a little more inconvenient than the previous post, since Warrant M'buke was still conversing with Ensign Judd, and the ensign took more than a normal allotment of space to begin with.

While this was going on, the reliefs for the four other stations entered and began trading pleasantries with those completing their shifts.

At 0745, when the XO entered, even though she wasn't due on the bridge until second watch, Ramirez had nearly had enough. He knew why everyone wanted to be on the bridge, but they were just about out of room.

Punctual as ever, the hatch opened again at precisely 0750 to admit Commodore Brighton.

"Commodore on the bridge!" Shuyler announced, more sharply than before. Everyone snapped upright, even those who should have kept their attention on their stations.

"At ease," the commanding officer directed. He looked around at the packed crowd on the bridge disapprovingly, and looked back to Ramirez as the officers resumed their places. He said nothing, though, and tried to maneuver himself out of the way. *Yargus* was never designed to carry anyone's flag, so it had no flag deck, nor any station for a flag officer on the main bridge.

The commodore's look was enough to finally exhaust the captain's patience. "All right, everyone not on duty, out! We can't work this way. Jenice," he added more quietly to Lieutenant Lewis, his second in command. "Set up the display in the mess to repeat from the scan board and let people know they can watch things from there."

She smiled. "Aye, Captain."

The hatch unlocked and people started filing out. "Contact!" CWO Eloise Kent shouted unexpectedly. The egress halted as everyone turned and flowed back in. Ramirez ignored them all; the surprise report driving all but immediate items into the background.

"Where away?" the captain asked.

"Dead ahead, sir. Course, speed, distance, everything is exactly as projected for *Pathfinder.*"

"Comm, signal to *Dagger* and pass on our scan readings. Scan, if they're where we expected them to be, how did we pick them up almost a half hour early?"

"Aye-aye, sir," WO3 Douglas Leonato, the comm officer of the watch, managed to squeeze in ahead of Kent's response.

"We've been working on the scan system, trying to increase sensitivity as much as we could these last few weeks. Our original estimates were calculated based on 100% efficiency. We must be running just above that now." There was an unmistakable undertone of pride in the warrant's voice, one which the captain decided was well deserved.

"Well done, Eloise. Extend my thanks to the rest of the department for me, will you?"

"Aye-aye, sir," she beamed.

"Any sign that they've picked us up yet?" the captain asked next.

"No change in their vector, sir," Kent replied. "They may not have ranged us yet, depending on their scan system."

"True. Let me know at once of any variation."

"Aye-aye, sir."

"Jenice," Ramirez called back to the undispersed throng, "keep everyone moving. We've got work to do here."

* * * * *

It turned out there really wasn't any work to be done, outside of the normal routine. The planning was already concluded, since nothing unexpected had been discovered.

Commodore Brighton did schedule a meeting for the fifth, twenty-four hours before operations were to begin. The main participants were Brighton, Ramirez, Chowdhury, Gvozdzius, and Roberts. He asked for Lieutenant Lewis and Second Lieutenant Ethan Malszycki, *Yargus'* head of security, to attend the conference as well.

It was a short meeting. The commodore took one sentence readiness reports from each section, and then announced that the primary attack plan, without any variation, would be launched at 0600 hours the following morning. Everyone said, "Aye-aye, sir," Brighton expressed his confidence in the Marines who would be doing the majority of the work, and the meeting was done.

Four minutes, twelve seconds, by Ramirez' chrono. That had to be some kind of record for a staff meeting. Not that there needed to be anything more than that, and Ramirez appreciated a CO who didn't drag meetings out to no purpose.

And there really was nothing else that needed adding. Everyone involved knew the plan backward and forward, after having weeks to hash it out, and the primary plan was extremely simple and direct.

Yargus' two attack shuttles, currently snugged into her ventral mounts, would be loaded with the main force of forty Marines. Their task would be to catch and board their target, subdue the thieves, and take control of her. *Yargus* would be providing fire support, and potentially extraction capability, if such were needed.

Major Chowdhury would be dividing Panther Dragoon into twenty two-man fire teams. Sheli Chowdhury and her wing would lead one platoon and

Vitek's team would head the other, each of them in one of the attack boats. The two shuttles were assigned different entry points, one near Engineering and one near the bridge, timed to gain access to their target at roughly the same time.

Chowdhury and Brighton had gone back and forth on how to handle the approach. Brighton wanted to hold fire on the ship until the shuttles were fired upon. His view was that if they were still here in the system, they must not have rebuilt an astrogation system which would allow them to calculate a jump vector. Or else they had not concluded repairs to the jump engine itself.

In either case, Brighton was certain that these items would be Leung's top priority to repair. If these had not been done, it was almost certain the lesser items, like weapons and scanners, would also not be ready to operate. Since Burton had reported that their eventual destination had been the Worth system and they were currently here rather than there, it could reasonably be concluded that they would not see them coming nor would they be able to fire at the approaching teams.

That was all well and good, as far as theories go, but the fact remained that the Marines would be running an avoidable risk in not targeting *Pathfinder's* weapons as soon as they could be ranged. Clearly, Chowdhury had been of a different mind on the matter, and had made her preference known during earlier planning sessions. To her credit, though, once Brighton had declared how things were to be, she had not uttered another word of dissent on the matter.

Ramirez was optimistic about the chances of the Marines making their approach without counterattack. Aligned as the ships were, the Marines would make their run in from dead astern of their objective. That was normally a blind spot on a warship, due to the interference of the propulsion system at the rear of the craft. *Pathfinder* wasn't actively driving her engines at the moment, but more of a ship's sensors, should they be active, and sensor duty officer's attention, would be focused on what was ahead rather than what was behind.

Perhaps a small thing, but Ramirez was happy for any advantage they could get. He wasn't counting on it, though. This was going to be his first hostile encounter; not just as a ship's commanding officer, but in any capacity; and he recognized his own inexperience. While still a novice at this sort of action, he was not an idiot, and only an idiot would think that the unexpected won't occur.

* * * * *

"Chowdhury to Bridge," the annunciator sounded late on third watch of 6 October, "we are buttoned up and ready to disengage."

"Talon Flight, Bridge," Ramirez responded immediately. "You are cleared for departure. All telltales are green. Good hunting."

"Thank you, Captain. Talon Flight clear."

"*Yargus* clear."

Ramirez noted the barely perceptible shiver of the two shuttles shoving themselves away, one after the other.

He could feel the tension of his crew as they came to the end of this ordeal. *Pathfinder* was finally there for the taking and they would have her back soon.

"Energy discharge, dead ahead!" WO2 Evans shouted from the scan console. Brighton shot to his feet from the temporary seat at the aft end of the bridge.

"Specify, Warrant!" Ramirez shouted right back. "Explosion, weapons, what?"

"Not weapons fire, Captain." Evans had control of her emotions now, and the report was much more professional sounding. "From the levels we're reading, *Pathfinder* has either opened a jump gate, or else she's just become a cloud of debris and plasma."

CHAPTER 26

6 October

Dagger

To: Admiral Conrad Cosina, CO RD-SpPrj/BuRnD/HQ
From: Commodore William Brighton, CO TF1/SQ4/1FLT
Time: 15:40:31 6Oct2787
RE: Progress Report

Admiral:

Operation Avatar commenced on schedule at 06:20, this date. Within ten minutes of deployment, an energy discharge in excess of 400 EW was detected on *Pathfinder*'s bearing. I cancelled the operation at once.

Further investigation has shown no evidence of debris or particulate matter in the area of space along *Pathfinder*'s course. This has led me to conclude that our target has managed to make repairs to the jump engines and has successfully transited out of the Antoc System. Our last communication from Sgt. Burton was accurate in all verifiable details and I assume, therefore, that the mission objectives will require us to head now to the Worth System.

Plans to incorporate a jump drive to *Dagger*'s systems are still a minimum of four weeks from implementation, and six to seven weeks from testing. Please refer to details reported in my progress report dated 10Sep2787. Repairs to *Yargus*' jump drive will be completed and tested within two weeks, two days ahead of the estimate provided on 18Sep2787.

Given the current situation, transferring my flag to *Yargus* may yet pay dividends. *Dagger* is better placed now to continue pursuit, since it is probable that they will be able to reach Worth by conventional means before a jump drive can be completed in her.

Captain Johnson will return to Antoc JP1 within three weeks' time, and will add her crew to the effort to complete the jump gate generator. When completed, she will retrieve the DaGaman prisoners taken at the asteroid base and transport them to Warner custody at Gateway, then continue on to the Worth system at best speed. I have reviewed with her contingency plans for her to join me there, but I do not expect to have any more available for the continued search than the resources of one ship. Considering the time necessary for *Dagger* to reach Antoc JP1, and the time required to completed the new gate generators, it is more than likely that *Yargus* will have concluded operations in Worth and left word for *Dagger* at Gateway prior to her arrival. Should *Pathfinder* have left Worth before I arrive, *Dagger* may again be better placed to offer pursuit.

If the pursuit leads me out of Worth space, I will send a report to you from there, and leave word for *Dagger* at Gateway. Barring a change of plans, I will send my next progress report on 21Oct2787 from Worth.

CHAPTER 27
7 October
WNS *Dagger*

With her nerves taut as a bowstring, Lieutenant Commander Fyonna Johnson sat in the padded command seat of *Dagger*'s bridge. An outside observer would have noted no immediate evidence of the cause of her tension. The officers around her, no enlisted personnel on this watch, were going about their tasks efficiently and easily. The bridge was a little on the silent side of normal, but that could be explained away by the fact that their destination was locked into the guidance system and there were no changes to be made, nor would be for weeks yet to come.

Her tension was not born of frantic activity, but rather its absence. She had nothing left to *do*, but plenty of thoughts to worry like a dog's well-chewed bone. There was the schism among portions of her crew, which had come to blows already. There was the order to break off active pursuit of *Pathfinder* in order to try to get ahead of her by heading for her purported destination instead. There was the ongoing effort to build a jump gate generator inside her ship. There was also, lurking unanswerable at the back of her mind, the memory of the explosion on *Yargus* when they had tested her jump engines. And, most nerve-racking of all, there were all those niggling little doubts about her suitability to be *Dagger*'s commanding officer.

Four months ago, she had been relatively comfortable in her role as Helm Officer on *Pathfinder*. A senior grade lieutenant and section head had been responsibilities that she had grown accustomed to in the two years she'd been assigned there. The shock of being physically removed from her quarters on *Pathfinder* was no greater than the jolt of being assigned as *Vanguard*'s executive officer. And now to be in command of her own ship! Worse, the commodore had removed himself from her ship, transferring his flag to *Yargus*. Now *she* was the ultimate authority. The weight of every decision seemed more than she could bear, and she had no one to whom she dared disclose her feelings.

She consciously willed herself to remain motionless as nervous energy fought to escape her somehow.

"Urgent message from *Yargus*, ma'am," Ensign Leslie interrupted her thoughts and caused her to jerk upright.

"Direct or buoy?" Johnson asked, then wished she could recall the words. Of course it would be via comm buoy. The commodore would be updating the Warner offices back on Earth at the same time. Standard procedure.

"I'll take it here, Ensign," she concluded, not allowing the junior officer any time to respond to her unnecessary question.

"Aye, ma'am. Transferring now."

Commodore Brighton's stern gaze filled the repeater screen at Johnson's station at the center of the bridge. His face was flushed with more emotion than she was used to seeing him portray, but otherwise the red-haired man looked the same as he always had.

"Captain Johnson," the message began, "I regret to inform you that *Pathfinder* is gone. She eluded our attempts to capture her before leaving the system. She was able to jump just as we began our assault. One bit of good news is that she was delayed long enough that we are not far behind her.

"Based on our current status, you will follow the directives outlined at our last meeting. Assume that your support will be needed in the Worth system, and make all haste to arrive. Do not assume, however, that *Yargus* will wait for you. The additional force represented by *Dagger* may be needed to apprehend Leung and the others, but I think that depriving them of time to move her again is more essential. They will likely feel themselves safe from pursuit now, unless they have managed to learn that Burton provided us with their destination. Even so, they are unlikely to remain in one place for a great deal of time.

If we are able to recapture *Pathfinder*, or if she has left the system before our arrival, I will send a priority message for you to await your arrival at Gateway. Given *Foundation*'s estimated time to completion for the gate generator system in Antoc, that is the more likely outcome.

"This comm buoy contains a copy of my report to Admiral Cosina, which outlines both my plans, and my expectations for you."

The commodore paused momentarily, and the stern and emotionless mask slipped away just a bit. One corner of his mouth tilted, and his ice-blue eyes warmed a few degrees.

"Good hunting, Fyonna. I hope to see you soon, but that will be up to you." Then his mask was back in place, and whatever thoughts and feelings her commanding officer might have were again inscrutable. "We *will* have our jump drive functional in two weeks. We will transit directly to Worth's only jump gate at that time. Brighton clear."

At first, Captain Johnson felt relieved at the report. In two weeks, the construction of a gate generator would barely have begun on her ship. Lieutenant Reed's latest report stated they did not even have the designs completed, since the matter surrounding the power supply systems had yet to be resolved. No one would expect her to be able to get to Worth that quickly.

Suddenly, the weight of the universe's expectations lifted from her and she found that she could breathe easier than before. Why was that? Her orders had not changed. Brighton's last instructions to her had been predicated on the possibility that had turned out to be fact; that *Pathfinder* would be able to jump before being captured. So why was the crushing weight of command suddenly easier?

With razor clarity, she saw what the difference was.

It wasn't that she knew any better now what her duty was than she had before. The difference was that she had been *told* what she needed to do. There had been someone there to look over her shoulder and make sure she was acting correctly. It was an illusion, though; an imaginary crutch. Brighton was not really in a position to monitor her activities and guarantee that her decisions measured up.

Nothing had really changed, only her perception of it. But she instinctively felt that something did have to change, or she would drive herself mad with second-guessing every choice. *She* had been given command of *Dagger*, and she had accepted that posting.

But she had never accepted the authority that went with it, not fully. She had always been looking for support from one source or another; Commodore Brighton, Lieutenant. Grant, a consensus opinion from her subordinates even.

That had to stop. She could see that now. Gathering recommendations was fine, but she needed to be the one to decide, and then to act.

"Lieutenant Corkill," she said, turning to the officer manning the Engineering board. Theresa Corkill turned sharply at the unexpected call.

"Captain?"

"Emergency power to engines."

"Ma'am?" the startled officer asked, certain that she hadn't heard right.

"Do you need the order repeated, Lieutenant?" Johnson asked, one eyebrow climbing into her black bangs.

"No, ma'am. I just… No, ma'am. Increasing thrust to 115% rated maximum, aye."

Several eyes looked as surreptitiously as they could manage at the captain. Her face bore the same mask that she had noted on Brighton's face a few minutes earlier, though she had not consciously willed it to be so.

"Astro," she directed, "assuming we hold this acceleration, with one hour at 240 gravities every four hours, what is our arrival time at the jump gate?"

Travelling in a straight line as they were, it was a simple calculation, and the Astrogation Officer, Lieutenant David Russell, had an answer for her in less than ten seconds. Two answers, actually: subjective time according to the clocks on the ship, and relative time according to the rest of the universe. They had already reached a significant fraction of light speed; enough that they had to keep accelerating just to maintain their velocity. The answer came back that such would put them at the gate six days sooner.

She spent several minutes in the command chair, silently working on the numbers herself, until she had the best plan she was going to be able to implement, balancing her desire to move faster with the need to keep the ship safe and the engines functional.

"Engineering, maintain this speed for five hours, then decrease to 255 gravities for thirty minutes before resuming."

"Aye-aye, ma'am. Maintaining two-niner-three gravities for five hours. Decreasing to two-five-five gravities for three-zero minutes, then resuming previous accel of two-niner-three."

"Very well. Ensign Leslie, attach orders to the comm buoy for *Foundation* and *Avram* apprising them of our new arrival time of 0210, 19 October. Instruct them to be standing by for gate transit at 0225. *Foundation* is further directed to provide us the means of transiting at once back into Betre system. Emergency priority. At our arrival, all Marines and prisoners should be aboard *Avram* or in shuttles and ready to board *Dagger*."

"Aye-aye, Captain. Transmitting. We should have confirmation from the buoy in…one hour, ten minutes, ma'am," the ensign acknowledged.

"Comm me when receipt is confirmed, Ensign. Lieutenant Russell, you have the bridge. I will be in Engineering with Lieutenant Reed."

"I have the Conn, aye, ma'am," Russell said by rote and moved to take station at the center console.

Johnson's surprise departure left everyone free to take their focus off their boards and look at each other. Several had raised eyebrows, but none were smiling.

"What was that all about?" Lieutenant Szoke asked from Helm, breaking the uneasy silence.

"I don't know," Russell said. "From what I could make out, she didn't receive any new orders from Commodore Brighton."

Lieutenant Amaya added, sotto voce, "And here I was thinking that Brighton was the source of fanatical obsession."

"It's not that bad, Pete," David said, not liking the shift in the mood of the bridge. "She knows that if we don't hurry, we'll miss the party, that's all."

"That's all?" incredulity colored the quick response of Lieutenant. Pedro Amaya-Garcia, this shift's weapons officer. He swiveled in his seat to bring his broad shoulders square to the OOD. "Have you ever before been in a ship pushing at fifteen percent over maximum? Have you even heard of anyone doing that before?"

"I have," Ensign Leslie piped up. She was physically the opposite of the dark and bear-like Amaya. At less than 150 cm and perhaps 45 kilos if she showered fully clothed, the petite blonde gave the impression of being something of a shrinking violet. That impression would only last until you heard her speak or looked in her eyes, and then her iron will would be clearly displayed.

"When was that?" the larger man boomed.

"Acceptance trials for *Katana*. We made it to 118% before we backed off," she replied, broadcasting her pride in the accomplishment, having spent her first tour in *Katana*'s engineering department.

"Uh-huh," Pedro grunted dubiously. "And how long were you at that speed?"

"Well, only for a minute, but I'm sure we could have maintained that speed a lot longer," Loren temporized.

"Five hours?" Pedro pressed.

The ensign did not respond, so Lieutenant Russell spoke up before the silence became oppressive. "That's enough, Pete. The Captain knows what she's about." You could hear the capitals in the title.

"If you say so, sir. I hope you're right, sir." There was more than a little mockery in Amaya's words. Both he and Russell had graduated from the Naval Academy in the same class, and both had made lieutenant on the same day. David was the superior officer at the moment simply because Captain Johnson had left him in charge when she headed to Engineering.

The captain in question hadn't made it that far yet. Coming out of the bridge hatch, she had realized that she had neglected to acknowledge her orders after their receipt. She was anxious to get to Lieutenant Reed and discuss how much she was going to push the engines before he decided that she'd lost her mind. She wasn't sure which was more likely to explode first, the engines or the engineer.

Lost in thought as she was, she probably would not have caught her omission if her path had not led her past the communications office. She headed straight in, as if that had been her objective all along.

Lieutenant Pol Strachowitz came out of his seat and to attention before her foot crossed the threshold. "Captain on deck!" The two other ratings in the room, neither of whom had names attached to their faces in her memory as yet, shot to their feet as well.

"As you were." she acknowledged. "Lieutenant, I need you to prep a comm buoy to send back to *Yargus*. How soon can you have it on its way?"

"Under ten minutes, ma'am. Would you like us to step out while you record?"

Why would he ask that? she wondered. Had she said anything to indicate that the message would not be for common consumption? Ah. He knows I have the bridge watch this shift. If I'm down here instead of up on the bridge, it must be, so he assumes, that I don't want the bridge staff to overhear.

There was nothing she was trying to hide, but it was easier to agree than explain. "Thank you, Lieutenant. I'll just take a few minutes."

"Aye, ma'am. We'll have the buoy ready to launch by the time you're done." With that he motioned the others and followed them out into the corridor, punching the privacy lock on his way out.

Johnson seated herself at the console and initiated the recording software. She recited the message headers, routing information, time of recording, all the mundane details that an official transmission would require.

"Commodore," she began once the essentials were completed, "your orders dated 6 October have been received and implemented. If it is at all possible, we will join you in the Worth System on 21 October to support you in the recovery of *Pathfinder*. If it is not possible, I will personally rewrite the laws of physics to make it possible.

"*Dagger* clear."

CHAPTER 28
7 October
Forrest Main Enclave, Earth

Amanda Forrest stood before the floor-to-ceiling mirror in her office's boudoir-petite, just a few steps away from where she customarily conducted business. The impression of power she strove to maintain required a certain look at all times, and so she kept the tools required for that maintenance close at hand.

She had selected a long, deep blue skirt that was a year or so behind the current fashion, but which was more comfortable than what the latest style dictated, along with a white undertunic and bright red overtunic. The colors also were too basic to be thought of as modern or fashionable, but they were colors that implied authority; and authority was also something comfortable to the CEO of the Forrest Family.

Once she was dressed to her satisfaction, she selected option three on the groomer, which gave her straight, shoulder length black hair, minimal shadow of a shade to enhance the green of her eyes, and red lips to match the overtunic. She made a final turn to left and right to assure that all was in place, then she stepped back into her office, to be met by her patiently waiting assistant, James. The two of them continued through that room and out the double doors into the hallway beyond.

James was quite young for the position he held; blond, of medium height, and slightly heavier than was considered healthy. None of those facts mattered in the slightest to Amanda. James had never forgotten anything that he had overheard, read, been told or commanded, nor had he ever failed to comply with any directions. Those were the qualities which made him invaluable to the extremely busy executive.

Arriving at her destination, she checked the time via her implant and found that she was two minutes early as she swung the door wide and headed for her seat at the head of the broad conference table within. There had been

147

a time, much earlier in her career, where she had deliberately been a little late to each meeting, to remind the others who was in charge. It wasn't long before she stopped such petty reminders. They never highlighted anything that anyone had forgotten, and they were simply inefficient.

She moved to the head of the table and was inwardly bemused to see every other seat already occupied. They all knew where the power rested, without any prompting, and had accommodated themselves to her schedule anyway.

Remaining standing, she took up the gavel, struck the block three times, and called the meeting to order. "The minutes of our previous meeting have been provided in the meeting's data packet. The Chair will entertain a motion to accept or modify these."

Noel Harkness, VP of External Sales, rose to his feet and stated in his clear tenor voice, "Madame Chairwoman, I move the minutes be approved as provided, and that the orders of the day be approved as outlined on the agenda." It was clear that Noel was impatient to get to the meat of their conference, not waiting to put the two actions into separate motions.

"Second," sounded immediately from multiple sources.

"It has been properly moved and seconded to approve the minutes and the orders of the day. Those in favor?" Many 'ayes' filled the air. "Opposed?" Silence. "Motion passed." Another tap of the gavel.

"The next order of business is division reporting. The Chair recognizes Penny." Amanda then took her seat while Penelope Forrest rose. Penny Forrest headed the Manufacturing Division, a position of considerable influence, but she was also the newest member of the Board of Directors, and thus the first to speak. Amanda had never been particularly close to Penny, a third cousin and ten years her junior, but found now that they were of the same mind on most things.

Not everything, of course. And not on the topic which would be central to today's…discussion.

"Thank you, Madame Chairwoman. As you can see from the summarized numbers on page seven, we reported a 14% decline in manufacturing for the month of September. Those, of course, are the reported numbers. If you'll tab to appendix C, you'll see the actual figures, which show that our manufacturing output increased by almost 8%. We plan t—"

"Point of information, Madame Chairwoman," Cynthia Forrest interrupted venomously. "Last quarter Ms. Forrest was projecting 20% growth for September. Why now the shortfall?"

Amanda was even more distantly related to Cynthia, but Cynthia and Penny had a long history. The CEO did not know the origins of the feud between the two, but it had never eased in the eight years that both had served on the board. "Ms. Forrest, would you care to respond?" the chairwoman directed more than inquired.

The lack of heat in Penny's response indicated that she had expected this question, and knew who would ask it. "Our expectations in the last quarter included the normal output from our three spaceship manufacturing centers; two of which are of Forrest construction, and one of which was acquired." Amanda noted offhandedly that this was a remarkably bland word to use to describe the violent attack that had resulted in the change of ownership. "From the first two facilities, the monthly output was about what was expected. However, the third, the Caspar Yards in the Worth system, were scheduled to complete three destroyers and two cruisers in September, but all five have been delayed for one reason or another. Those ships should complete shortly, and our numbers will come back up at that point."

"What were the reasons for the slip in the production schedule?" Amanda pressed with deceptively calm features.

Penny looked as if she'd bitten into something sour. With some reluctance, she said, "The staff of that facility is primarily made up of the same people who worked there prior to our acquisition. In recent months, there have been some signs of resistance on Granada, and a portion of that appears to have translated into slower work being performed in the shipyards. This behavior is being corrected now, and should not affect future production."

Amanda Forrest rose to her feet then, jaw muscles bunched in controlled anger. Penny shrank back into her seat, yielding the floor without a quibble.

"Understand me, all of you," she said in flinty tones, "this plan has been held together by the barest of threads from the very beginning. You all know that I urged you to find other means of providing a secure financial future for our corporation. My main argument at the time was that this course is an all or nothing proposition, with too many breakpoints where it could fall apart.

"We are gambling with our lives! We are gambling with everything Forrest holds and represents! There is far, far too much at stake for us to sit back and accept any excuses. Our planning and scheduling absolutely depended on the fighting strength we have been promised, when it was promised." Amanda's gaze bored in on every board member. Penny could not meet it. "We cannot accept business as usual on this endeavor. All of the Board needs to be on the same page on this."

"Human Resources," she nodded to Julia Bentsen, who headed that division, "has arranged for the training and staffing of those ships. Now we have almost 900 people sitting around waiting for a ship to board. Internal Sales," a nod to Serina Forrest, " has arranged the stores to lade into those ships, which are now sitting somewhere taking up warehouse space.

"I was opposed to this strategy at the outset, but now that we have begun, we have no options but to make it work. Delays like these are poison to our plan!"

She fixed Cynthia with an icy gaze. "And trying to play politics among ourselves is even worse. If we keep digging pits for each other, we're all going to wind up stumbling into them."

Cynthia pretended the jab was meant for someone else, and looked without shame from person to person, all innocence. Penny had the decency to be embarrassed, and looked unswervingly at her shoes. Amanda realized once again that her role was becoming more and more that of babysitter rather than potentate. She resumed her seat, but her fiery gaze stilled all other movement in the room. Finally, in an impatient voice, Amanda said, "Was that all, Penny?"

The addressed individual rose sheepishly. "The overview is spelled out in the précis, and the details are in the attached appendices. If there are no other questions, I have nothing further."

"Very well, let's move on to the report from OPA. The Chair grants the floor to Mat."

Mathias Kinsey-Forrest stood erect and smiled as all eyes turned to him. He enjoyed being at the center of things, and the current direction the corporation was heading had placed him in that position on an ever more frequent basis. It had been his knowledge of that fact which had been the catalyst in his developing this course of action and urging its implementation. When it succeeded, people would remember who had the foresight to develop it. Amanda would not be long in the Chair by then. He tried with only limited success to keep a smile from his face.

"Unfortunately, Madame Chairwoman, there is little change in the difficulties being handled by Off-Planet Affairs. We've had great success in training officers and building up both a Navy and a Marine Corps in the last ten years. I am confident that, given three more years, four at the outside, we will have a military force superior to Combined Fleet in both quantity and quality.

"A major portion of our planning for the past two years has involved the military advantages granted by point-to-point transportation from onboard jump point generators, and all construction during that span has left extra space in the engineering sections to allow an easy retrofit once we had the design specifications in hand.

"That should have happened no later than 1 July, according to the arrangements we had made. Our last communication with our contractors indicated that all was proceeding on schedule, but we still have not received the schematics and designs we need. I fear, at this point, that we may never get our hands on that information.

"We still have not been able to determine exactly what happened, though we do have one contact inside Warner that may be able to acquire the debrief records from Brighton and his crew. Even that may not be sufficient to clarify the situation, however, because it is likely that they were expelled from the ship without knowing anything of Leung's plans."

He stopped then, his dark eyes dancing from face to face and reveling silently in their rapt attention. "My suspicion, though it is based only on circumstantial evidence, is that DaGama Aerospatiale was able to find a way to contact Leung directly and outbid us for delivery of *Pathfinder*. That Family was equally informed of the stakes for which we were playing. When we made our own, far superior, contact on the project and were able to cut them out of the deal, they had to at least suspect the reasons why. If they were able to identify our inside party, they may have made a preemptive offer and thus cut us out."

Amanda thought Mat was probably right. His theory explained all the known facts, at least, not that there were a great deal of those. She tapped a note into her datapad to feel out Paola DaGama on the subject, and forwarded a copy to James. That call was going to have to be handled delicately, to say the least. Communication was not at all cordial these days; but business was business, and all the Families accepted that and played by the same lack of rules. Perhaps she could yet make an arrangement to purchase the new technology. She would have to; too much depended on it.

"In any case," Mat was continuing, "without the advantages of a jump drive, we will need additional time to build up and train a greater force. Unfortunately, this is not likely to be possible either."

If anyone's attention had been wandering, as Amanda admitted that hers had, it was brought sharply back at that bombshell of a statement.

"Point of information, Madame Chairwoman!" Noel burst out as he rose. Even angry, he followed Robert's Rules and did not address Mat directly. "Would the Vice President for Off-Planet Affairs care to explain the meaning of his last statement?"

Amanda glared. She hated Mat's grandstanding, and Noel was so predictably blunt, he never failed to rise to Mat's bait. "Mr. Kinsey-Forrest?" She used the extra formality to let him know of her displeasure. He responded with another ill-concealed smile.

"As the Board is aware, part of my area of responsibility is to monitor activity in other Families that might interfere with our plans. In most cases, we have only been able to obtain peripheral information, but we have enough that certain patterns emerge to show us the overall picture. So, while we do not have details or schedules, we do know, or can deduce, several things.

"First, we know Captain Brighton spent only a few days in Sol system before leaving again. We know that at least four ships were retasked to accompany him. From the speed of his departure, it must be surmised that he intends to track down *Pathfinder*. If my earlier suppositions are correct, this pursuit can lead them nowhere but to the DaGama Family.

"This would not be enough to keep us in the clear, however." Many faces that had relaxed at the thought of Warner blaming DaGama for their troubles tensed once again. Kinsey-Forrest continued, "DaGama is aware of our activity, certainly, as well as the other Families that we have contacted and obtained cooperation up to this point. There may not be anything we can do to keep them from finding out about the entire operation, and that discovery will certainly come before we are prepared to handle it."

There was stunned silence for several heartbeats as those seated absorbed the facts, and the attached consequences. The years of planning and preparation, the money spent on those preparations, all of it could be lost if they were attacked before they were ready.

"And what do you recommend we do about it, Mat?" Julia spoke for the first time. Amanda ignored the lapse in protocol, as she wanted to know the same thing. Not because she thought what he had to say would solve all the Family's problems, but because she knew his proposal would involve an expansion of his power, which she would need to curtail.

It was a familiar pattern, as old as mankind, to use scare tactics to gain adherents and followers to your own cause. His answer, though, caught her off guard.

"I don't know," Mat stated flatly. This was unexpected, and Amanda immediately wondered what he was playing at. "I have given this matter a lot of thought over the past few weeks, and I confess that I have not been able to see a solution. I bring it now to the Board in the hopes that one may be found."

With that, he seated himself and a buzz of low conversations filled the room. The CEO rose to reclaim the floor, and the murmuring eased, but did not disappear completely.

"Madame Chairwoman," Penny Forrest said, erupting from her chair as if burned.

"The Chair recognizes Manufacturing." It was another sign of Amanda's current disfavor, not using the woman's name.

"If the Chair would indulge me, I would like to outline a potential course of action before making a formal motion to enact it," she said.

Amanda nodded and resumed her seat.

Penny cleared her throat, organizing her thoughts as quickly as she could. The idea she was about to share had only just occurred to her, but she needed to fill in all the details before speaking, or she would lose even more status within the group.

"The course we have been following for the last several years has focused mainly on creating our own military force capable of compelling Warner and others out of their monopolistic position over exploration and shipping areas. To that end, we have been constructing ships and weapons in secret facilities so that we could have sufficient might before anyone was aware of our actions and tried to stop us."

"We know all this already," Noel snapped. "Mat just informed us that we're going to lose that particular race!"

Penny flinched away from the harsh words. She had hoped for more time to fit all the pieces of her idea together and probe for gaps, but she had to jump in now, it appeared.

"All right, but maybe we can win the race by moving the finish line," Penny countered.

Amanda sat up straighter in her seat. "In what way?"

"So, if we were to attack Warner now, we would inflict heavy casualties on them while they were surprised and unprepared, but they would quickly stiffen their lines and call on Combined Fleet and the other three Families with ships, which would overwhelm us.

"The same can be said if we stay on our current track and Warner discovers what we're up to before we are ready for them. We will have lost the element of surprise, and they would still be able to call on CF and the others for backup.

"But it would be a different story if Warner were isolated from the others, if, instead of being able to call on Combined Fleet they instead had to defend against them," her finger stabbed the table with each succeeding point, "and us, and Fermi, and Portales, and Sterling."

A dozen sets of eyes lost focus as her words painted them a picture they didn't mind looking at. "The way to go about this is to use our resources in the media and in other Families to subtly get people thinking that Warner is becoming another Parkinson." Parkinson's name being mentioned caused more than a few nervous looks. They were the Family which had started the Interstellar War at the beginning of the century.

Penny continued, warming up to her topic now. "There are several indisputable facts that we can point to as the basis for suspicion. They have the largest private navy, just like Parkinson did. They have high tariffs for exporting to their worlds, just like Parkinson. They are expanding to new worlds faster than any other Family. Yes, it's mostly luck that the branch from Earth which they claimed wound up containing more jump points than anyone else had, but the man in the street wouldn't see it that way unless someone pointed it out, and I see no reason to do so.

"From that base, we can fabricate more 'facts' to support the view we want people to have. The most important point will be to convince the Ruling Council. The RC controls Combined Fleet. With them on our side we will be stronger than Warner, even if all the other Family Fleets joined with them."

Penny could see many thoughtful faces, a few grinning, but most importantly, none hostile. Not even Cynthia. She smiled inwardly.

"Therefore, Madame Chairwoman, I move that the Forrest Family adopt a strategy to isolate the Warner Family from its current allies by spreading suspicion about their past, current, and possible future actions."

"Second," sounded immediately from all eleven of the other Board members.

CHAPTER 29
8 October
Yargus

Commodore William Brighton stood near the aft bulkhead of WNS *Yargus'* bridge and dispassionately analyzed the particulate scan being displayed on the forward screen. It showed what it had for the last two days; *Pathfinder* was gone. They had completed an exhaustive search of the surrounding area, but it had turned up nothing. His ship was gone, slipping from his reach just as he was about to reclaim her.

There was no anger showing on his features, and truly, very little remaining behind his well-schooled mask. Rage had flared up inside when the energy discharge of a jump gate being opened was first detected, but he had had time to master it since. It had surprised him that he had been able to rein in his feelings as easily as he had. Perhaps age was bringing more patience with it. Or maybe he just didn't have the energy to be fueled by his passions anymore.

That was the more likely reason. Eight weeks of plentiful food and mild exercise had recouped somewhat the dropped mass experienced as a result of his chilling flight across this very system, but he still found that he did not have the necessary energy to work, or even sit and think, for long periods of time as had always been his characteristic method. Sudden weariness enveloped him again, both physical and emotional in nature, and he had to fight not to let a yawn escape him.

"Commodore," Lieutenant Shuyler interrupted his thoughts, "scan is complete. No evidence of debris. Looks like she's gone, sir."

"Yes, Lieutenant, I am forced to concur with that analysis. Turn us about and reduce our velocity, if you would," the commodore directed.

"Aye, sir. Helm, bring us about on reciprocal. Maintain acceleration," the astrogator ordered. As it was third watch, the middle of the night, Norman Shuyler was the Officer of the Deck at the moment.

"Mr. Shuyler, I'm heading to my quarters. Would you let Ensign Roberts and Lieutenant Burkhalter know that I would like a meeting with them at 0800?"

"Aye, sir. Orders logged, sir," the OOD responded with one of his customary short answers.

Brighton left, deep in thought, and let his body carry him to his bed without conscious direction. He could now abandon at least half of the strategies that had been filed away in his normally agile mind for possible future retrieval, and he was sorting through them and doing so.

Arriving at his hatch, he entered his key and passed through. A small portion of his brain guided his body through the necessary items of disrobing and such while the majority concentrated on what he would need to discuss with the chief engineer in a few hours. A general plan, with several ancillary items formed itself, and he stored this in one corner of his thoughts as he reclined himself.

He was unconscious before he was completely horizontal.

* * * * *

Roberts had been dismissed earlier than normal the night before, having the entire evening to herself after dinner. So she had done what she had dreamed about doing every day since being forced off *Pathfinder* and living for months without being out of arm's reach of several people—several unwashed and entirely too fragrant people.

She took a long, hot bath.

This singular event was not so easily arranged as might be expected. The head adjoining the junior officers' wardroom where she bunked did not have anything but showers; nor were any of the others onboard equipped any differently. There was only one place on the ship where you could find a tub, and that was in the medical suite.

Access had been arranged through that oldest of military traditions, barter. When Doctor Agee had cornered her and loaded strict instructions on her that she was to make sure the commodore followed her orders to consume at least 2800 calories each day, she had traded her cooperation for access to the soaking tub.

Once she had permission to use it, though, she found that she had no time to do so. Commodore Brighton had rapidly regained his former energetic state and he seemed to be on the go at least eighteen hours of every day. And it was very bad form for the staff ensign to retire for the evening before her flag officer. Such things tended to find their way into a junior officer's personnel jacket, almost guaranteeing said officer being passed over for promotion. Ensign Roberts had set a goal to be the first of her class to make lieutenant, and while one of her classmates had lucked into beating her to that rank, she was sure that Samuels's promotion would be found null and void in the end, having been granted by a traitor. That meant there was no way Ms. Jherri Roberts would call it a day before Commodore Brighton did.

Unless said flag officer should happen to tell you to turn in early, which was what he had done at 1700. "It's almost certain that we won't find anything, Ensign," Commodore Brighton had told her. "There's no sense in both of us wasting our time waiting for the inevitable to be confirmed. Why don't you turn in early tonight, and I'll see you in the morning."

"Aye, sir!" she had responded enthusiastically. Eating a larger than normal dinner, she had convinced herself that not going straight to bed would not be disobeying an order, so long as she was in bed earlier than her normal hour. That wasn't hard to accomplish considering Brighton's tendency to still be up at the commencement of third watch, and it allowed plenty of time for a nice, long, relaxing soak in the tub.

After dinner, she returned to her berth to rid herself of various bits of equipment and her uniform, donning her exercise outfit and filling a duty bag with her toiletries and a towel. She was back out of the hatch in under three minutes.

The bath had been so relaxing that she hadn't gotten to enjoy it all. She awoke in tepid water and returned to her bunk.

* * * * *

Roberts' alarm sounded at 0500, but it was Ensign Ragnarsdottir's launched pillow at 0503 that woke her up. "Sorry, Jessica," she mumbled while keying off the offending device and rising. Once upright and moving, she felt amazingly rested; energetic and mentally sharp. Donning her hastily discarded exercise gear, she headed aft, through the reinforced bulkhead and into the boat bay.

The voluminous area was very crowded, as it always was these days. Available space had been sharply diminished by the influx of all the Marines now on board. Housing them all had been solved by the expedient of breaking out their field gear and pitching tents in precise rows all along the port half of the expansive bay.

The starboard side had been turned into the Marine's exercise and training yard, and it was to this section that Roberts aimed her strides. A bull-necked sergeant had Marines lined up in ranks and was leading them in calisthenics. Roberts quickly dropped her gear against the wall and moved into an open space between PFC Allen and PFC Lance. They smiled in welcome and she smiled in return, though she felt like she was standing in a hole, she had to crane her neck so far to look them in the eyes. She waited for the cycle to return to standing, and then she joined in with eight count pushups.

The silent welcome and acceptance of the Marines was a comfort to her, one she hadn't really expected to feel. Months ago on Antoc-B3, now named Fujinami after a fallen comrade, Roberts had started exercising and training with Major Chowdhury. At the time, she thought it had been an attempt to keep from being caught unprepared again. To some extent that was true, but there were other reasons which were much deeper, and it wasn't until the major had sat her down in a lunchroom on Gemmill and talked things out with her that she had realized it.

The memory was only a few months old, still fresh in her mind. During the time when everyone was working to repair and provision the Portales ship for a return to Earth, Roberts and Chowdhury had found themselves alone together in the lunchroom of the hospital where many were still recovering.

Roberts had been more than a little uncomfortable in that situation. The major had an intimidating air that she carried with her everywhere. Add to that her opinion of ensigns as the military's lowest form of life, if they even qualified as life and not just wasted environmental capacity, only added to the intimidation for Roberts. When Chowdhury approached and asked to join her, however, there was no polite way to decline.

"I haven't seen you at calisthenics for a few days, Ensign," Chowdhury commented lightly while taking a seat.

"Captain has asked me to be his liaison officer, and several meetings have conflicted lately. Don't worry, I haven't given up on my training," Jherri responded, a little defensively.

"Is the exercise helping?" The tone with which this question was asked gave Roberts the impression that it had a heavier meaning than what it appeared. Confused, Jherri looked up and noted the intent look in the older woman's eyes. They reminded her of Captain Brighton's in that instant. Not in color or shape, but in the way they could look into your soul to gauge what was there.

"I'm not sure I follow your meaning, Major," she had ventured warily.

"You feel guilty about Le Vesconte's death." This was not a question, but a statement. It also seemed to Jherri to be a completely different topic, as if Chowdhury's train of thought had jumped its tracks.

"Huh?" she began. "No, I—" Chowdhury's raised hand stopped her short.

"I know, Ensign. Believe me, I know. In your head, you understand that what happened was not your fault. But inside, that's not what you *feel*. You saw him die right in front of you and you keep berating yourself for not saving him. Your friends die and you survive because you were, by sheerest chance, in the right place to escape the hand of death. One minute you're part of a full company of friends, and the next you're responsible for a handful of survivors—" She stopped herself abruptly, realizing she'd begun dredging up her own ghosts. She looked into her lap, suddenly embarrassed.

"Believe me, I know," she repeated in a quiet voice. Roberts suddenly understood why Chowdhury and Commodore Brighton seemed to be able to look inside you and see what others couldn't. They had been there, and done their own soul-searching. They could understand you because they had come to understand themselves.

Chowdhury cleared her throat, and her voice was more normal when she spoke again. "So let me tell you what I think is going on with you, if you don't mind." Jherri was surprised to see that the Marine was actually waiting for a response. Roberts nodded and Sheli continued, "You recognize that what happened to Drew was not your fault, but you still feel guilty. These two facts war inside you, until you start to rethink what happened. Maybe if you'd recognized what was going on sooner, you could have kept Drew out of the line of fire. Maybe you could have changed places at the table with him. Maybe, if you hadn't been as concerned for your own safety, you could have broken from cover and dragged him to you sooner. Maybe he would have lived with quicker assistance. Or maybe you could have stopped all of us from going with the DaGamans at all.

"And the more you repeat those points to yourself, the more reasonable, the more real they seem. Understand that a person's subconscious cannot differentiate between truth and lie. That is a purely conscious function. So whatever you tell yourself, you believe. When your subconscious accepts your criticisms as real, it makes you feel worse about Drew's death. Eventually, you convince yourself that you were completely to blame for his having been killed. All the responsibility for that hole in his chest falls on your shoulders, and you need to find a way to punish yourself.

"Exercise is a popular option in the military. You push your body to the point of exhaustion and pain, because you think you deserve it for letting a comrade get killed. Perhaps you justify it by telling yourself that you just want to be more prepared if the situation ever repeats. Deep down, you know the reason why, though."

Chowdhury's intense gaze never released her, and she was unable to hide the fact that, if she were truly honest with herself, every word described her exactly. At the time she had begun working out with and learning to fight from Johnson and Chowdhury, she'd really had no business doing so. She had already lost more weight than was healthy, and while food was a little more plentiful on B3 than at any other time during the nightmare escape from the Antoc system, adding unnecessary physical exertion was just short of suicidal. Perhaps, a tiny voice inside her said, that had been her wish.

Silently, she began crying, her shoulders shaking from the effort to keep the sounds inside. Chowdhury made no move to console the ensign, but neither did she have any apparent negative reactions. She sat still and waited for the junior officer to recover her equanimity. It was a considerable wait.

Finally, Jherri asked, "What should I do then? I *do* feel I'm to blame for his death."

"You shouldn't," the Marine countered sharply. "I was responsible for Drew's death. I was responsible for everyone's security. I still am."

"That's ridiculous! You didn't know—"

Chowdhury's angry and menacing look stopped Roberts short, her throat suddenly too dry to speak.

"I *did* know. Not the specifics of what was coming, but I knew the people leading us to their camp were not what they appeared to be. But with them already all around us, I couldn't warn everyone without them noticing."

Chowdhury watched the girl's eyes go wide in shock at her words and then narrow in anger. "You snake!" she shouted, shooting to her feet. The fist she sent flying at the Marine's head did not even make it halfway there before Chowdhury's hand caught it and stopped it short. Jherri swung with her left, but the older woman effortlessly pulled the captured right fist into the path of the other. Roberts cried out in pain and was pushed back into her seat. Chowdhury had never even gotten out of her chair.

"You finished?"

"I suppose," the girl harrumphed, nursing her hand and pride both.

Chowdhury smiled. It might have been the first time Roberts had seen the Marine do so. "Well that's a step in the right direction, then. You've shifted the blame from you to me, and you understand how easy it is to do. Now we just need to put it back where it belongs."

"So you mean that you didn't know. You were just playing with me?"

Chowdhury's countenance became grave once again. "I did know," she asserted. "At least, I knew enough to be suspicious. And I was not playing with you, just...demonstrating how people's minds work."

Roberts glared. "Then you *are* responsible for Drew's death."

Chowdhury sighed. "Ensigns," she mumbled under her breath. "Ensign Roberts," she began, as if explaining to a child, "there are a list of factors twelve clicks long that contributed to his being killed two months ago, going all the way back to when his mother and father met. Any variation in that chain might have led to a different outcome. All of those are relative. Let's look at the situation from a legal standpoint, because at least then we have some solid rules to judge guilt by."

Jherri nodded suspiciously and Sheli continued, "All right. What was the cause of death?"

"A rifle shot."

"What kind of rifle?" the Marine pressed.

"I don't know, an energy weapon. I think it was plasma discharge rather than laser, because I saw the flash." Roberts gulped after replaying the moment in her mind.

"Good. It was a Sikorskiy GLG-20 high-energy plasma sniper rifle. Not one I would use, by the way. A trained ear can hear a high pitched whine while the weapon recharges. Was the man who fired the rifle acting in self-defense?"

"What? No! Of course not. Drew wasn't armed, and none of us even knew he was there. We were no threat to him."

"Was he acting to protect others from imminent harm?"

"No."

"Did a state of war exist between the Warner and DaGama Families at the time of the attack?"

"No, but there certainly is justification for a war now!"

"I agree, though I hope it can be avoided. Now, lacking any legally recognized defense for his actions, would you say that the man holding that GLG-20 was guilty of murder?"

"Yes. Without a doubt."

"And you conspired with him to make sure he killed Drew, is that correct?"

Roberts looked horrified at the thought, and found herself unable to respond.

"Right," Chowdhury said gently. "Now that we have established who the guilty party is, the only way that you or I could share in his guilt would be if we had been working with him from the start. Clearly, that wasn't the case. But the more you take that guilt on yourself, the more you take it away from where it really belongs. Can you see that now?"

Roberts was a long time answering. The tears had started once again, but this time she did not try to hold them back.

When the emotions were washed out of her and she had command of her voice once again, Jherri found herself thanking the wiser Marine, both for talking her through her grief and for saving her life on A3, which she realized she had not done before.

Chowdhury got a funny look in her eyes and said nothing for a long time. Finally, she said, "Ensign, a long time ago, I was in that seat thanking a man who came back to save me. So I will tell you the same thing he told me. 'Just use your brain from now on. Always know what is going on around you. Always know what could happen before you act. And always, always, always have a way out.' That advice has been seared into my brain ever since that day and it has saved my life at least four times. I hope it serves you as well."

Roberts knew instinctively that the man she referred to was Captain Brighton, but she had never been able to bring herself to ask the major to tell her the story behind those comments. After the brief exchange on Gemmill, Chowdhury had again drawn back from any social contact with the young naval officer. Once again, she had been just another ensign to the Marine, as far as any evidence could support.

Jherri had been a little hurt by this. Though she continued working out with the Marines, Chowdhury never joined in, nor did they exchange more than a few words thereafter. She had begun to believe that she must have said something wrong. She was considering her complicated relationship with all of the other officers and crewmembers who had been aboard *Vanguard* with her when she noticed one of the Marines moving toward her tentatively as the exercise formations broke up and everyone began moving to their individual training. She waited and watched as Corporal Jay came to a point in front of her and waited for her approach.

He came to attention and gave her a salute, something almost unheard of considering the internecine territorialism of the two branches of the military. "Ma'am, could I have a word?"

Roberts returned the salute, bewildered. "Certainly, Corporal. What is it?"

"Meaning no disrespect to the ensign, but could you tell us in what unit you saw combat? I do not doubt Major Chowdhury's word, mind you, but you don't look to be old enough to have served in any battle that we've ever heard of." He looked like he desperately wanted her to believe that he did not doubt his commanding officer, and he wouldn't be there at all if the small group had not been nudging him along.

The noncom's explanation did nothing to clear up the ensign's bewilderment. Trying to stall for time while she sought to understand, she asked, "What did Major Chowdhury say, exactly, that you do not doubt?"

"Well, Riley here," he indicated the brown-haired young woman to his left, "overheard the major leave instructions with Sergeant Kucinich to allow you to train with us whenever you wanted. She said, 'That woman is a combat veteran. She watched my six in a tight spot, so you treat her like a Marine, Top. I consider her family.'"

Roberts had a hard time keeping her face impassive. She was sure that she could not have done a very good job of it as unbidden tears threatened to fill her eyes, but her voice was steady when she replied.

"The major was referring to the Battle of Le Vesconte, though we also saw action together in the Battle of Fujinami. You can look up the details when they are no longer classified. Until then, I cannot comment on them myself. If you'll excuse me."

The memory of the stunned look on all their faces brought a smile to hers as she moved from the general exercise area to one corner, where she would again get banged and bruised learning how to fight like a Marine.

CHAPTER 30
8 October
Yargus

The message from the duty officer about her meeting with Commodore Brighton and the Chief Engineer was waiting for Ensign Roberts as she returned to her room at 0745. Jessica was absent, probably breakfasting, and Tomiko, the other female ensign on *Yargus*, would not be off duty until watch change at 0800, so she had the quarters to herself.

She peeled out of her damp clothing and dropped it directly into the laundry chute. A quick shower allowed the overworked muscles of her arms and legs to ease their tension, but wasn't so relaxing that she felt tired when she stepped out; the extra sleep from the night before helped there. Drying off, she could identify only one new bruise earned that day, although it was larger than most. Hand-to-hand combat training with Marines was definitely full-speed, full-contact, and no excuses.

She pulled yesterday's undress uniform out of the laundry return slot and dressed quickly and efficiently. A final look in the mirror while she brushed her hair and collected it back into a ponytail showed her ready for duty. She recovered the datapad and other items she would need and affixed them in their accustomed places before quitting the room.

The walk to the Commodore's quarters was not a long one; forward on her level, up two decks, then forward down corridor two almost to the bridge. Sprinting, it was even shorter. It was precisely 0755 when she tapped smartly on the Commodore's hatch.

Rather than permission to enter, the commodore stepped out and acknowledged her presence with a brief nod before heading aft. Roberts hurried to keep up, staying a half step behind and to one side. Her datapad was already in her hand, for she knew what was coming next. Commodore Brighton's mind didn't seem to shut off while he was asleep; it just queued up orders that came spilling out as soon as he was conscious.

"Both forward missile rooms are slow reloading," he said without preamble. "I know it's not my ship, and I shouldn't notice, so would you discretely mention to the XO that a few extra drills might be in order?"

"Aye, sir." Her stylus added this item to her ever-growing list.

"Unless *Dagger* has put more distance between us than I would expect, we should have heard back from Commander Johnson forty minutes ago. Check with Communications after our meeting with Lieutenant Burkhalter and make sure they haven't missed anything."

Another few strokes added this, though she expected that word would arrive before she had the opportunity to investigate. It wasn't likely that the comm duty officer would have neglected to forward a message to the commodore, any more that it was likely that Commander Johnson would forget to respond.

Roberts was sure there were more orders stacked up and ready to spill out, but Brighton's long strides meant they would remain unvoiced for a bit longer.

Arriving at the bulkhead to the engineering section, the commodore paused and looked at the young officer for the first time that morning. She was still making notes to herself, and Brighton waited for her to stop and look up. "Did you sleep well, Ensign?"

"Yes, thank you, sir." She was surprised at this show of concern by her commanding officer. He generally paid no attention to how she was holding up under the load he placed on her, nor had he been solicitous of her comfort before this.

"Good," he harrumphed, looking away. He straightened his uniform tunic and stepped through without any further comment.

Jherri did not have the opportunity to puzzle out the change in behavior. When they stepped through the hatch into the engineering section it was an anthill of chaotic activity. It was all she could do to stay out of the way of crewmen crossing her path and attempt to keep up with her superior. Brighton moved swiftly and unerringly to the Chief Engineer's portside office. He looked at his chrono as he came to a stop and rapped sharply on the metal hatch at precisely 0800.

There was no answer.

After a moment, Brighton's jaw muscles worked, and he bounced his fist off the door resoundingly. Still nothing.

He turned away quickly and moved deeper into the engineering section, glancing down each intersecting walkway for Burkhalter or the shift lead. He was still searching when a crewman from the comm room found him instead.

"Message from *Dagger* for you, sir," he said while saluting. The thin sheet was handed over, and he saluted again before retreating. Brighton's glower, though not meant for him, might have occasioned his swift departure.

Brighton did not even notice Crewman Corrain's haste, as he read the brief note. He smiled when he finished, an unusual enough reaction that made Roberts wonder what Commander Johnson had said.

"Ensign, we need to have a staff meeting. Please make the arrangements, and see to it that Lieutenant Burkhalter is in attendance. We'll also need the captain and XO, department heads, and Major Chowdhury and her second."

"Aye, sir. What time would the commodore prefer?"

Brighton looked at his chrono and said, "0830."

"Aye, sir," she responded, silently panicking at only being allowed nineteen minutes to organize the gathering. "With your permission, sir."

He dismissed her with a wave, and she managed to wait until he was out of sight before increasing her speed to a run.

* * * * *

Lieutenant Shanarra Burkhalter was not in her seat at 0830, but she was the only one missing. Jherri assumed Commodore Brighton would want to start on time so she stood just as the row of zeroes appeared on the chronometer. Brighton waved her back down to her seat. She promptly sat, astonished.

Captain Ramirez glanced around the silent table, but no one seemed anxious to speak. "Commodore," he began, with what he hoped was a firm voice, "we may need to begin without Lieutenant Burkhalter. She is supervising the repair of the number five coolant pump."

Brighton studied the captain for several agonizing moments before finally responding. "Very well, Captain."

When Roberts began to rise once again, he raised a hand. "Ensign, if you don't mind, I'll just take the floor."

He did not wait for any response, which made everyone more comfortable, being more in line with what they expected from him.

"I received word this morning from Captain Johnson. She does not expect to complete a working jump engine in the near future, but she does hope to arrive at the Worth system as much as ten days earlier than we had originally planned. I realized at once that, firstly, I should have expected as much from Captain Johnson, and secondly, that I would need to revise my tentative plans for retaking *Pathfinder*.

"Following that line of logic, I found that I had made my plans without any input from this group. While it is ultimately my responsibility to decide on a course of action, it would be unwise of me not to also enlist your help in outlining how to handle whatever we may find once we transit into Worth. That said, I will open the floor first to suggestions as to what the situation might be at our destination, and then what our options are to deal with those eventualities."

165

"Well, Commodore," Jenice Lewis, the ship's executive officer began, "There are really only two possibilities I can see. Either we follow close enough behind *Pathfinder* that she's still in the system, or we're too late and she's gone."

A look flashed briefly across Brighton's face that might have been disgust or merely disappointment, but he quickly blanked his features. No one at the table had missed the expression, however, and all knew what his opinion was. "Yes, Lieutenant, what you say is true, but entirely incomplete. For instance, if the ship is still in Worth, it could be before they have met with their contact or after. The Forresters could have a cargo ship or a war ship as the contact. They could have multiple ships. They could be anywhere in the system. We may be alone in the system, or joined by *Dagger*. If they've gone from the Worth system, how do we find out what their destination truly was? I want us to start thinking about these possibilities now, so that we have options planned in advance for every possible eventuality that we can identify. Once Lieutenant Burkhalter is available to join us, I hope that we'll have a better idea how much time we will have available for planning.

"Having said that, I will not jump after *Pathfinder* while we are still unprepared for what we may find," he said with grave finality. "I do not like surprises."

"Commodore," Lieutenant Shuyler offered, "I don't know if we can plan for every possible situation. There's too much we don't know. How can we plan without any data?"

"Lieutenant, we can plan without data because that is what is required of us. No commander ever has all the facts in advance. The universe will rarely make things so convenient for anyone. While it will be difficult to map out several different plans, most of which we will never use, wisely employing the time we are forced to spend here, before we jump, will give us the greatest chance of success."

Lieutenant Burkhalter entered before anyone else could comment, taking the last remaining seat. She hadn't taken the time to change out of her overalls, and she did her best to keep her greasy hands out of sight under the table.

Brighton nodded to her, but made no comment, waiting instead for ideas from the others. Shuyler leaned her way and quietly brought her up to speed.

"Well, anything is possible, but I think that we will still have a lot of elements in our favor," Ramirez began. "Since Worth is a Warner system, we will have the cooperation of the military base and the gate control station. If they are still in that system, we can quickly identify the Forrester ship and stop them from jumping out. If they have gone, we can back trace them from the scan logs and departure logs and find out where they have gone, and when."

"I agree with the captain," Lewis added. "I don't know why they were dumb enough to pick Worth as the rendezvous point, but it certainly plays into our hands."

"She's right, Commodore," Shuyler added. "They've painted themselves into a corner. We jump in two weeks," he paused and looked at Burkhalter, who nodded agreement, "and there's nothing we'll see that we can't handle with or without *Dagger* to back us up. I mean, how bad could it be?"

"How bad could it be?" Chowdhury spoke for the first time, her contralto voice devoid of the scorn her eyes held. "Forrest has invaded the system with overwhelming force. The orbital military station has been atomized. All life on Granada has been destroyed. The jump gate is currently in enemy hands. There are Forresters in control of the shipworks and mining fields, building warships to attack other systems. They've taken the time to invest the jump point with additional defenses; mines, weapons platforms, battle drones. The plans for *Pathfinder*-type engines were transmitted months ago and each of their ships can jump clear of us without warning," She paused in her litany of possible disasters to scan the shocked faces around the conference table. "That's how bad it *could* be," she said when the silence had stretched to an uncomfortable level.

"You say that by choosing Worth, Forrest has made a tactical error. I say that with as much planning as we have already seen them use, a mistake of that level is unlikely. It is more likely that they feel safe there, and if you Pollyannas poke your heads in there with this level of overconfidence, they will waste no time chopping them off and handing them to you."

She might just as well have thrown a dead body on the table, from the stunned silence in the room. Clearly, those were not possibilities which anyone, except Commodore Brighton, had even remotely considered. Chowdhury scanned each face at the table, using a look Roberts had thought was reserved for ensigns.

"No one who goes into battle planning for only the best outcome ever comes out ahead, and usually does not come out alive. If you swabbies can't get your limited intellect wrapped around that idea, this is going to be a painfully short operation."

CHAPTER 31

9 October
Earth – Forrest Main Compound

A long, silver mag train floated a bare centimeter above its metallic pad as it sped through the green grass and ordered rows of productive farmland. The area would once have had an identity; Cherokee Nation, North Carolina, Asheville; but such names, and the ownership they implied, belonged to the past. The name of Smoky Mountains still had meaning, and still described the area flashing past the passengers' windows, but no Family owned this land, and no Family protected it.

This was part of the wild world outside the Families' Enclaves that got by the best it could with what it had. This region seemed to be doing well, Colonel Stefan Valencia thought, but that was logical given its proximity to Forrest's main Enclave in Raleigh. Raleigh's population of twenty-eight million needed food to survive, more than could be produced within its own walls. Forrest produced the best farming and heavy equipment to be had, which the independent farmers and ranchers needed in order to make a living. Mutually profitable trade was a natural result.

Valencia smiled to himself. As a senior officer of Warner's military intelligence bureau, he constantly had information crossing his desk which he needed to jigsaw into a coherent picture. He preferred situations that made logical sense, and the occasions in which logical situations presented themselves to him were so rare that they must be relished. Generally, a situation had to be puzzled out, probed from every angle, and throttled into submission before the logic underlying the reality coalesced.

The adolescent girl across the aisle continued to stare at him, and the appearance of the smile caused her to yank on her mother's sleeve for both her attention and her protection from the scary man. The woman shushed her absently and went back into her doze.

Valencia did not appear as he normally did, and hadn't since boarding the train in Houston an hour before. He currently had ash blond hair instead of his natural black, cut much shorter than usual, and he was dressed eccentrically in loud colors. The guitar case in his lap, when added to the clothing, labeled him immediately as a musician, and that was what he intended; a convenient pigeon hole for people to place him in before promptly forgetting they had ever seen him. When you cannot be easily categorized, people tend to study you and to remember you.

As always, Stefan's facial features presented the most difficult problem. His beak-like nose was larger than normal, and so he had deliberately brought attention to it by connecting a nose stud by a silver chain to a similarly-colored ear plate. Again, human psychology guided his choice. He could not hide his nose without surgery, which he would not do unless it was necessary, so he instead showed the world what it expected of a flamboyant performer. Had he tried to conceal his nose, or minimize its dominance of his face, it would have remarked more comment, and would stick in people's memories.

His current persona was the fifth he had adopted since a suborbital up to Denver from Quito the day before, each transition taking place completely out of anyone's sight or notice. The guitar case contained, in addition to the expected instrument, his next several identities, each just as simple to categorize as his current persona.

Valencia had been studying sheet music and ignoring everyone around him throughout the trip, but his head came up when the car darkened momentarily, then brightened a bit with an artificial light. The train had left the unprotected zone and entered the kilometers-broad Forrest compound itself. His stop was only a minute away. The other passengers responded to the visual trigger and started putting away loose items and gathering their burdens. Stefan did the same, blending in with the rest. When the vehicle stopped, he rose with the others, shuffled along with them, and queued up at the customs gate.

"Name?" the seated woman in Forrest green prompted while he handed over his ID and travel documents.

"Black Adder," he responded gravelly.

The vacant, pig-like eyes continued their bored study of his papers. It took several seconds of processing before she registered that the name given did not match what she was reading. Finally, she looked up, comparing the holographic image to the physical one. Again, there were notable discrepancies, and a lengthy pause indicated a few lethargic synapses were trying to resolve the paradox. It turned out to be too much for her.

"Glenn," she called over her shoulder without rising. A competent-looking man stepped out of the guard kiosk and strode purposefully over. *Ex-military*, Valencia categorized immediately, *probably near retirement*.

"What's the trouble, Elena?" his clear tenor voice asked, while his eyes never left Valencia.

"His name don't match his papers," she whined.

"Name?" Glenn asked warily.

"Hugo Parkinson," Valencia said. Elena glared.

"That's *not* what he said," she pointed out.

"No, and my apologies for that, pretty lady." Her demeanor improved markedly. "Usually, I get asked to repeat myself, and it gives me an opportunity to invite people to my performances." With that, Stefan pulled out a dozen low-quality holoflimsies announcing that Black Adder, a bright new luminary in the Retrosynth Jazz world, would be performing for two nights only at Molly's.

"No harm done," Glenn said after looking over the advertisement. "But remember that security is not something to play games with."

"Yes, sir," he responded., sounding suitably chastened.

Glenn went back to the kiosk without looking back, and two more minutes with Elena had his paperwork completed. He offered her a flimsy again, but she declined. That was good, since he wouldn't be going anywhere near Molly's while he was here. If all went as planned, he would be back out of the unoriginally named Enclave One before his first imaginary show.

Stefan took the slidewalk heading east, toward Molly's, in case anyone was watching. At the next junction, though, he switched to a high-capacity radial line headed for the central hub. He dropped off before reaching Central Park, and then jumped up nineteen levels, circled partway around the hub, and then headed back outward for almost a kilometer. By then he was certain no one was following him and he finally headed into a public restroom, one of the large ones with multiple exits. There was no surveillance inside, and he made sure he was far enough in to be out of range of the corridor cameras before selecting a stall and tabbing the opacity field on.

When Valencia emerged, Black Adder was nowhere to be found. The last of that person was a few grams of ash rolled up in protective sheeting that Valencia dropped into the waste as he exited. The colonel was now dressed as a well-to-do businessman answering to the name of Carmichael. His suit was an understated charcoal and purple, perhaps five years out of date, with broad lapels on the tunic and pants tight through the legs. Valencia favored this style because there was almost no limit to the kinds of things you could carry unnoticed beneath the lapels. Anything he needed to protect himself if it came to a fight could be stored there. His carry bag held little more than his next identity. He was down to just two remaining.

Stefan checked himself into The Courtyard, a business-class hotel, without any difficulties. His reservation was three weeks old, confirmed that morning. He wished the desk attendant a pleasant day and took the lift up to the fortieth floor, went down the hall to room 40612, entered, and dropped his bag on the bed.

Checking his chrono, he saw he still had nearly an hour before the next step in his plan. Extra time needed to be built into the schedule, to account for the unexpected. He had been somewhat perturbed that other concerns had dictated that the margin was that close. That he hadn't needed to worry this time would not keep him from worrying just as much next time, of course.

He turned on the holoset, then, using its covering noise, spent seven minutes sweeping the room for surveillance equipment. He found none, but he would still assume they were there. A few more minutes used up in the "settling in" process, which businessmen accustomed to travel exercise with practiced proficiency. If things went wrong and his room were searched later, it would appear no different than thousands of others in the hotel.

Valencia considered what else he might do productively with his time, but nothing came to mind. Not one to enjoy waiting around, he grabbed his bag and left the room, deciding to conserve the remaining extra time in case of future need. His next eight steps did not need to coordinate temporally with anything else, after all. For the ninth, however, timing would be everything.

He paused in the hall and turned back to pull the door closed and ensure that it locked itself. The lift carried him quickly down to the lobby, where he asked the desk attendant if she might recommend a good restaurant that served Americano food. She named three, and asked if he would like her to get him a reservation. He consented, and wound up headed for Henry's Rustic, up five levels and in one ring.

Dinner turned out to be surprisingly good. Henry's knew how to do a steak right, and the vegetables were fresh and tasty. Valencia did not hurry his meal, and studied business news on his tablet as he ate. He indulged in dessert, since he had lots of extra time, apple pie a la mode.

He continued to sit in his seat, reading the article while scanning the room, for several minutes after the bill was settled, but not long enough to be unusual. Valencia gathered his things and left without hurrying, and wandered aimlessly for the next hour.

Eventually, he made his way into Central Park, and made use of some of the dense foliage to make his next identity switch. After the weapons and tools had made their way into his oversized pockets, the gray and purple suit became another few grams of ash. He couldn't depend on the ventilation in such a large area to disperse them for him, so he sealed them into a packet for later disposal.

He moved far enough through the brush to be out of range of any surveillance that might have seen him enter, then joined a slow-moving stream of people meandering along with no particular urgency. Valencia used his time to go over his plan yet again, still looking for potential flaws. He knew from sad experience that this was the part of an operation that got more agents killed than any other; the part where everything was going smoothly, and you felt like things were almost over.

It was a hard thing to stay on guard all the time without looking out of place. Too many relaxed when they thought there was no need to be vigilant, then found out, too late, that they were wrong.

Stefan had been in this line of work for a long time, and it was mostly due to his first trainer. In Valencia's first live assignment, his training officer, Lieutenant Nagarao, had been swept up only twenty meters away from him, because she hadn't noticed subtle clues around her. No one had ever seen or heard of her since. Valencia couldn't think of a single thing that the lieutenant had taught him in the five months they'd worked together, but that final lesson was just as fresh today as it ever had been: Never let your guard down.

Valencia whiled away his extra time by walking all over the complex, pausing frequently to admire various displays, to read a billboard, to listen to a street performer. He noticed everything, and from what he could see, he wasn't being followed in person. There were still cameras everywhere, and there was no way to determine whether or not he was being tracked that way.

If they were onto him, and tracking his movements, they would have to move in to pick him up, and he was watching for that.

Changing identities in the public restrooms was intended to foil video surveillance, since privacy laws kept cameras out of that area, and so Valencia ducked into another one, and came out the other side as someone else, having disposed on his previous identity along with the packet containing the one before that. This would be the last change, and not carrying an extra person's clothes and identification around with him made him that much less likely to be caught. He acted as if he were in a bit of a hurry now, and checked his watch frequently, making the change in apparent personality with the ease of long practice. To any casual observer, he was just another laborer trying to get to work before his shift started. Valencia was certain no one was following him, nor giving him any more than a passing glance.

Still, he kept himself vigilant. Overconfidence was not going to be a problem; paranoia was too much a part of his soul for that.

He caught a lift at the next junction, and got off on the ninth floor of the gamma level, then started back south. It was twelve blocks more before he caught another lift all the way up to eta level, and then a slidewalk that took him west for three kilometers. He'd had to step off the moving platform with everyone else and pass through the scanner and show identification when he entered the government sector. He handed his card to the attentive guard, who found nothing out of the ordinary and handed it back.

Valencia's next stop was not far; a locker room, where he again changed his appearance by donning a service tech uniform. The real owner of the uniform was enjoying a free night out courtesy of a winning raffle ticket. He would never know it had not been a legitimate contest. As soon as he was presentable, Valencia was out into the corridor, moving down two levels to the maintenance office where he punched in at the clock with everyone else and collected his orders for the day.

The first two orders were of no particular importance to him, except as a way of remaining inconspicuous. He spent an hour and a half replacing a section of ventilation ducting that had seen better days, and over two hours rewiring an office thermostat before he moved on to the item he was interested in.

At precisely 12:15, he walked through the front doors of the Forrest Family main offices. There was another security checkpoint here, with armed guards highly visible plus automated defenses that only a trained eye would notice. The stakes kept getting higher as he went, but Valencia allowed no sign of his nervousness to show. The one thing that was the same here as at the two previous security stations was the lack of interest he aroused from anyone.

Within fifteen minutes, he came out the other end of the gauntlet and took a tube up to the top level. When he stepped off at the correct floor, a Marine security sergeant was waiting for him. That was not expected, but again, Stefan's features betrayed nothing.

"You Ericksen?" the guard asked, more bored than menacing.

"Yeah."

"Come with me." The Marine indicated the corridor to Stefan's left. The office Valencia needed to enter to complete his mission was to the right. The Marine, Verchou, his name patch read, did not lead the way, nor did he follow behind with his weapon at the ready. The man kept pace with Valencia a meter distant, with his blast rifle slung over the far shoulder.

Stefan tried to determine whether or not "Ericksen" should be curious, and finally decided that it was likely enough that he would risk asking. "Is something wrong?"

Verchou didn't slow, but turned to study Valencia. "Why?"

"Well, I've never had an armed escort while I worked before." Which was true for "Ericksen," at least.

"Must not've worked in the executive wing before."

"Nope. You do this with everyone that comes in?"

"Every person, every time. Except for board members, of course."

"Of course," Valencia agreed readily. The conversation had provided a bit of distraction to the guard, enough that he didn't notice Valencia dropping a small bit of melted plastic and fused metal into a waste can as they walked by. His fingers hurt from the heat that had bled through as the device had turned itself into a bit of inert and unrecognizable junk.

Valencia walked into the security office to which Verchou motioned so they could carefully look him over. For the first time in a dozen years, Valencia was going to have to resort to plan B.

CHAPTER 32
9 October
Dagger

"I tell you, it won't work," Warrant Officer Stephen Long said to his companion.

"Of course it will. You're just an idiot and can't figure out how to do it," Senior Chief Durrant replied pugnaciously. "Grant wouldn't have told us to do it if it couldn't be done. You're probably not smart enough. Maybe I should go get a monkey, instead."

Long held his anger in check, again, but just barely. He knew the Senior Chief was goading him. He was just trying to make Long react. He could always fall back on military courtesy. Long was a Warrant Officer and so, technically above any enlisted personnel, regardless of how high they had risen or their seniority. This was not always the case in practice, however.

Long looked around the conference room, where they were having their planning meeting, looking for any sort of distraction that would keep him from tearing into the Senior Chief seated next to him. The room was small, with eight chairs surrounding an oval table. The walls were painted a soothing pale blue, but the color choice had no effect on Long. He was still recovering from his months-long ordeal aboard *Vanguard*. His strength was returning slowly, but he still felt tired most of the time. Still, that was not the reason he was allowing Durrant to say things to him that no one else had ever survived uttering.

Long finally looked at the stocky non-com seated on the chair beside him and gritted his teeth. Being forced to work with the bullying Senior Chief was the most difficult thing he had ever had to do; which was saying something, when you considered that he had spent the last several months adrift on a lifeboat, powering the crew's escape by peddling a generator. Starvation and depravation were child's play compared to the restraint it took not to cave in the teeth of this blowhard. Every fiber of the Warrant Officer's body and soul

175

cried out to tear the idiot to pieces. If he hadn't been ordered by Captain Johnson to get along, he would already have done so. More than once.

Maybe he could do it quietly, he thought to himself. *If no one could tie the deed to him, then he hadn't really violated the order, had he?* No, that was not the way Long worked. He would do it openly or not at all.

"The figures don't work. You can't make the generator put out enough power," Long said finally. "The engines on *Pathfinder*, use giant capacitors to augment the direct output, but even then, look at the energy requirement for a field with a radius 2.7 times bigger. We're talking about almost twenty times the power. I'm telling you, the figures don't work."

"So you are just going to give up and slither away like the worm that you are."

Long's arm shot out so quickly that Durrant never saw the fist that knocked him from the chair.

His grin grew even larger as the burley Senior Chief picked himself up off of the floor and leapt to the attack.

* * * * *

Ensign Josiah Mitchell threw himself exhaustedly onto his unmade bunk. He hadn't had time to make it for the last three days; in fact he hadn't had time to do more than change his uniform and shower for the last several weeks. The fact that his other three bunkmates were just as harried and overworked as he was himself, was no consolation. All the ensigns were pushing as hard as possible to complete their normal duties as well as the extra assignments they had been given by the captain.

"Hey, Jos, where have you been? I thought we were supposed to go shooting this afternoon," Ensign Hayes said merrily from his overflowing desk. His lamp was highlighting several open hard-copy technical manuals and his own handwritten notes.

"Humph, very funny," Mitchell said into his pillow as he shut down the overhead lights. "How do you have any time to go shooting? This engine project has me running sixteen different directions. Not to mention my regular duties."

"I don't know, maybe I'm just better at managing my time than you."

Mitchell launched his pillow at his friend. "Yeah, *clearly* that must be the answer. You were *so* good at managing your time at the academy. Or have you forgotten I had to coach you through your finals because you were still studying for your mid-terms?"

"That was all part of my time management plan. You got everything correlated and that saved me time. See, time management." Hayes returned Mitchell's pillow with much more force than he had received it.

"Hey, just let me sleep, will you?"

Any response was interrupted by a knock on the door.

"Argh, see if you can manage to get that, will you," Mitchell grunted to his friend as he rolled over and faced the darkness of the bulkhead.

Hayes chuckled to himself and finished making another note on the chaotic notepad in front of him and rose to key open the hatch.

"Ha," he grunted with a laugh as he left the visitors standing in the corridor and went back to his desk, "it's for you, Jos."

Mitchell rolled over and blearily looked to the hatch. He briefly considered rolling back and ignoring the pair, but finally sat up and dry washed his face with his hands, trying to erase the fatigue and despair from his features.

Mitchell took in the state of his visitors as he walked slowly to the hatch. "Warrant Long," he said as he moved to the hatch, "have you two been fighting?" Mitchell heard a snort from the direction of Hayes' desk but didn't turn to verify the smug look he knew he would see on his friend's face. *Maybe I should rethink the definition of friend*, he thought to himself.

"No, sir. We were ordered not to do so as a condition of our sentencing. Why would you think we had been fighting, sir?"

Mitchell studied the cuts and bruises that covered both faces in front of him as well as WO Long's blackened left eye and Durrant's obviously broken and bleeding nose and decided he was just too tired to deal with any of this.

"If that is the case, Warrant Long, what can I do for you two... um... gentlemen this evening?"

"It's the numbers, sir. They don't work. We can't figure how to generate enough power to make the engines work."

"Did you try…"

"Begging your pardon, sir," Long interrupted, "we've tried everything. You're welcome to give it a try. Um… Sir," he said as he handed over the stack of papers he carried.

"Ok, see me at 1000 tomorrow, like we scheduled," Mitchell said as he closed the hatch in their bloody faces.

* * * * *

Mitchell studied the papers in front of him but he wasn't really seeing them anymore. He had expected to find some simple arithmetic error in the other men's work, but he had found none. In the six hours since then, he could find no errors in their equations either. He didn't know what to do. He had been given the task of supplying the power to make the engines work.

He knew the theory was sound, *Pathfinder's* engines had worked perfectly and these engines were exactly the same, only scaled up to fit the larger hull of *Dagger*. In the face of this, he had done the calculations now eight separate times and gotten the same result each time. Math and Astrogation had been his best subjects at the academy. He no longer doubted the results. The power requirements grew exponentially and it was just not possible to power the engines enough to create the jump bubble needed to get *Dagger* through the jump.

He was a failure. Not only that, but his failure would doom the entire project. What should he do?

He let his head fall forward to slam onto his desk.

"That's the answer, you just pound your head until it starts working right," Hayes called from his bunk where Mitchell had thought him asleep.

"I can't make the numbers come out right," Mitchell said without raising his head.

"You want me to take a crack at it," Hayes said with a chuckle.

"Only if I want two plus two to suddenly become six."

"How about Friedman. He'd take a look if you asked," Hayes said, referring to one of their missing bunkmates.

"Soloman's pulling a double in engineering. I think they were going to try to get the new relays installed tonight. He doesn't have time to look into this."

"Are you really sure it doesn't work?" said Hayes with a serious voice.

"Yes."

"Then you know what you have to do now."

"What's that?"

"Remember what Brighton said?"

"Which time?"

"Be sure, then act."

"Oh, crap."

"That's right."

"I've got to tell Reed."

"You got it in one," Hayes said with a laugh. "Turn off that light on your way out."

* * * * *

Lieutenant Thomas Reed was angry. Those who didn't know him well would not have been able to tell, but most of his engineering crew had been around him long enough to be able to interpret the signs. They discretely backed off and left the offending crewman to his own devices.

"Did I just hear you correctly, Crewman Vargua?" the chief engineer asked quietly. "Did you just say that it didn't matter if you did your job correctly?"

"No, sir. I was just saying to Tommy that the job was done."

"Don't lie to me crewman. I heard what you said," he said taking another step into the crewman's personal space.

"No, sir. The job is done and up to spec."

"Are you calling me a liar, Vargua?"

"No, sir," the crewman said, standing at attention and remaining silent after his direct answer.

Ensign Mitchell walked up behind the chief engineer but, wisely, also remained silent.

"I'll tell you what you are going to do, Crewman. You are going to unseal that connection, strip it down to the bare wiring, re-splice the wires and then reseal it. You will get Warrant Yong to verify and sign off on each and every step. Furthermore, you will not go off shift until it is complete. Is that clear?"

"Yes, sir."

Reed stood looking at the crewman as if waiting for him to make some complaint. Their shift was scheduled to end at 0700, two hours from now, and he had just been assigned to do six hours worth of work. Vargua remained silent and Reed began to move on. It was good to remind them that all work needed to be done up to standards.

As Reed turned to leave, he almost ran down the new ensign.

"What are you doing in my engine room, Mitchell?"

"Sir, I ran into a glitch in the power flow calculations and I'd appreciate it if you would take a look."

"I've got enough work to do without doing yours for you. The calculations are there, all you have to do is build the wiring. Now if you'll get back to work, so will I," Reed said as he started to move past the ensign.

"I beg your pardon, sir, but the calculations are incorrect," Mitchell said as he moved back in front of the engineer.

"You, in all of your infinite wisdom and experience, are going to tell me my job?" Reed sputtered, the fury rising in his body and his face starting to turn red. "Get out of my engine room and do your job."

Mitchell took an involuntary step backward then stiffened. "Sir," he said as loudly and clearly as he could, "I have checked the calculations multiple times, they are incorrect. The engines will not work."

"I have checked the calculations," Reed said, his voice rising in intensity, "there is nothing wrong with the calculations. Now take yourself out of my engine room, do your work and leave me alone."

Mitchell took another involuntary step back as the chief engineer pushed past him toward his office. He glanced around the engine room where the crewmen were conspicuously hiding the fact that they were listening to every word of this confrontation.

Mitchell took a deep breath and took the irretrievable step that now seemed to be his only course of action. "Sir, I disagree with your assessment and feel that your order is not in the best interests of this ship. Therefore, I respectfully notify you that I cannot follow your order and will be forced to appeal your decision to the Executive Officer."

Mitchell watched Reed freeze in the doorway to his office and heard a slight gasp behind him. Reed did not turn around or acknowledge his word in any way other than to stop in place. Mitchell knew, as well as Reed and all of the crew, that what he had just done wouldn't end the career of either officer and it was technically allowed by the regulations but it just wasn't done. There were only two justifications for disobeying a direct order from a superior officer. The first was the belief that the order was an illegal order and the second was that the order was not in the best interests of the ship or the fleet. Mitchell knew that the order he had just received from Reed was not illegal, only wrong, so he had been very careful in the wording of his statement. Still, having made the claim, he could no longer discuss this with the Chief Engineer until the matter had been resolved. Reed stepped into his office and slammed his datapad down on his desk rather than escorting the Ensign to the XO as the regulations mandated, so Mitchell was left to himself to go and try to find Lieutenant Grant.

CHAPTER 33
9 October
Earth – Forrest Main Compound

Colonel Valencia took the toolbox back from Verchou, who had remained within a couple meters of him throughout the process, and sat down to put his shoes back on. The scan and search he'd had to endure were thorough in the extreme. Had he not disposed of the electronic surveillance device he'd brought with him, he was certain it would have been discovered, and he would have been arrested and questioned; another process Valencia was sure would have been thorough in the extreme.

Shoes sealed, tools collected, back-up plan in place, Valencia rose and followed the course indicated by his guard to the server room. Again, Verchou neither led nor followed and remained just out of easy reach. It was both pleasant and frustrating to see the security force taking their jobs seriously. He liked to see people do their jobs well, but *his* job would be significantly easier if these particular guards were a little on the lax side. Fortunately, Stefan was not one to count on adversaries to do what was convenient for him.

After several turns and hundreds of meters, passing offices and meeting rooms to either side, they arrived at their destination. Verchou set his left hand on the biometer plate, which checked everything from basal temperature to DNA. The algorithms found him to match closely enough to their golden records to let him in, and the door clicked audibly as the actuators released the locks. The Forrester pushed the door open and held it for "Ericksen" to precede him in.

Valencia stepped into the room and was assailed by the sound of the ventilation system pumping air from the ceiling, whirring past his ears and passing through gratings in the floor. He could see only one workstation among the aisles full of hardware racks, so he headed straight for it and sat down. Verchou followed him.

"You staying for the whole job?" the spy asked.

"Yeah," the guard answered. Just like that, Plan B was busted.

"Okay, hand me the software you need me to install," Valencia said, without any pause to indicate there was any disappointment at having to move down his list of backup plans.

"What software?"

Valencia turned and looked at the guard. "Are you kidding me? No, of course you're not. Look, the software that gets installed back here has to go through as much scrutiny as I did, probably more, so the Data Systems Office sent it down here last week for you guys to run it through all your tests before I got here to add it to the core. They told me it would be here waiting."

"There wasn't anything about that in today's passdown."

"Can you check again?" Valencia prompted.

"Sure," Verchou agreed. He didn't move. "You'll have to come, too. Can't leave you here alone."

And that was it for Plan C.

Valencia and Verchou retraced their steps back to the security office, where Verchou positioned himself near his charge while they checked for the new security programs "Ericksen's work order told him to install. There was no sign of the actual data stick, but one had been logged in the receipt book from the DSO on the third. The security work logs showed the normal screenings had been started, but there was no end time or results recorded.

"I'm sorry about this, Ericksen," Verchou said earnestly, "but it looks like we'll have to set this up again for next week some time."

"Not your fault, man. I understand. But listen, does that mean I'm going to have to go through that poking and prodding again next time?"

"Yes." Verchou's response was so flat, Valencia was sure round two would be every bit as thorough as the first time. The newly-formed Plan D died an abortive death.

"In that case, what if I could get another copy of the programs sent over. How long would it take to get them cleared?"

Verchou looked for confirmation to his superior, who seemed more engrossed in a game of three-handed pinochle; deceptively so, since he responded immediately. "Three hours."

Valencia pulled out his netpad to check the time. "I think I can make that work," he said, while pulling pieces of Plan E into place. He dialed a number into his pad from memory and waited for a response.

"Adrian...I'm up here in the executive wing for that install you assigned me. Thanks for the warning about the security screen, by the way...Yeah, you're hilarious...Don't think I'll forget...What? Is this about the fifty I owe you? I said I'd get it to you on payday...I suppose you arranged for the software to go missing, too?...No, I am not kidding...Look, forget that, how long will it take to put the data on a new stick and have it messengered over?...Yeah, I know you're busy...It's not *my* fault the install pack went missing…All right, so an hour and a half?…Right…And the new service pack for #19 just released this week, so you should include it, and I'll update that at the same time…Of course I'm thinking ahead. Someone in our department has to…I wasn't implying anything, what were you inferring?…Ha, I thought so! See you when I'm done here." Valencia keyed the disconnect tab and put the pad back in its sheath.

"My boss says he can have it over here in an hour and a half, which probably means two hours; three hours for the scan to give it the all clear, and three more to install it all," Valencia reported.

"That's gonna make for a long day for you," the guard leader, a Marine corporal, noted.

Valencia grinned at him. "I get paid by the hour, and overtime starts at four o'clock."

"I guess we know how he's going to pay off his $50 debt," one of the other guards said. "And since he's got some time to kill, maybe he'd like a chance to earn a little extra. You play pinochle?"

"Sure. What stakes?" Valencia agreed warily.

"Dollar a point with short numbers. I doubt you'll win or lose too much," the leader said. The grin he attempted to conceal indicated the opposite.

Valencia had hoped that the relaxed atmosphere of the security office would lead to some kind of opening, but that didn't look likely to happen. Verchou never asked to join the game, and a place was never offered him. He sat behind Valencia's left shoulder, where he had a clear view of the directional hologram showing Valencia's hand and the playing table. As ever, Verchou remained just outside of easy reach, yet close enough to respond instantly to any false move. There was no chance to even propose a Plan F to himself, but Plan E was still viable.

They cut cards for partners, and Valencia wound up with Private Isom. Three hands were enough to determine that Isom and the other private, Gilchrist, were both rather conservative in their bidding, while Corporal Escrima was daring without being reckless. Escrima had taken the bid five times, losing one, before Valencia had a hand worth bidding on. Isom passed the cards Valencia needed and they more than made their bid, but Gilchrist had laid down a double pinochle in meld, which put them out – and then some. After one game, Valencia was down $90.

There was a break in the action then, while the three other card players went out and completed their rounds. Verchou remained in the office with "Erickson," and Valencia was able to get a few pieces of conversation out of him before the others returned.

The next round, Valencia was paired with Gilchrist, who successfully shot the moon with nine hearts, making for a short game, and Valencia coming out $20 ahead on the day.

The luck of the draw put him back with Isom for the third game. The cards were more balanced this time, and the lead went back and forth three times before Valencia barely made his last bid and went out $15 ahead of Escrima and Gilchrist.

It was well past 1600 before the last hand ended, but the new security team waited patiently until the game was done to receive their passdown information. It appeared to be an oft-repeated scenario, from the way everyone treated the situation. The passdown included "Erickson," of course; explaining why he was there, why he was waiting, and what to do with him. The data stick arrived during the passdown, and the NCO in charge of the oncoming shift took it and left the guard room. Valencia sent his prayers out the door with him that his contact had understood all of the encoded instructions in their conversation, *especially* variation #19.

Debts were settled before the guards left. Valencia had a modest $35 gain, while Gilchrist had been the big winner. No one seemed overly upset at the outcome, and Gilchrist offered to buy the first round on his way out the door.

The oncoming shift was headed by a sergeant instead of a corporal. She clearly wouldn't have passed the yearly physical fitness tests she'd have faced in the WSMC with forty extra kilos. As short as she was, she'd have to grow twenty centimeters in order to be square. Once the others had left, Sergeant Cavalos went into the security office and shut the door, not to be seen again. The other three members of the team, Breckenridge, Latham, and Jones, found various tasks at individual desks; essentially ignoring Valencia. If he still had his electronic bug, he might have tried to slip out and plant it. Since he didn't, Valencia found an unoccupied seat and settled in to wait.

It was less than half an hour before the datastick returned from whatever initial scanning it had undergone. Latham was their data specialist. He took the stick, loaded it into his terminal, and started running a set of secondary testing programs. After about twenty-five minutes, he pulled the stick out and handed it to Valencia. "All clean," Latham said.

"I applaud your efficiency. The corporal said it would be three hours," Valencia said, accepting the datastick and putting it in his sleeve pocket.

"Well, yeah," Latham explained, "if you run one test after another. I just kicked them all off in their own process stream. Quicker that way."

Valencia gathered his things and went to the door. No one met him there. "So, I know my way to the server room, but one of you is going to have to let me in."

"He means you, Latham," Breckenridge said.

"Not I," Latham fired back. "I did the software check. That means it's your turn, Sylvia."

"All right," Breckenridge sighed, closing out her terminal and getting up. "But that means Jones gets the 1700 rounds. Let's go, Ericksen."

Valencia followed behind the brunette, back down the same halls he had passed through before. There was more activity than there had been earlier; office workers trying to finish up their day's work and head for home.

Breckenridge provided the biometric data and the door unlocked. Valencia went in and sat at the single terminal. He put the datastick in the port and waited for it to autodisplay the file list. He turned discretely to see where Breckenridge would position herself, only to find that she had not come in with him!

You could tell what sort of leaders the two NCOs were, based on how much effort their troops put forth. Verchou had been on the ball throughout their time together, but it was clear that Escrima expected exactly that sort of professionalism. Cavalos' expectations were more modest, and probably did not extend much beyond being left alone.

Not one to waste an opportunity, Valencia got straight to work. His first step was to get the install started in its own window, and then to pause it. With that set to cover the eventual return of security, he wrote a quick script to grab every 19th character from the 19th file displayed on the list. This produced a note from "Adrian," who did not actually work in Data Systems, nor for Forrest.

The message contained instructions for reassembling the needed code from bits and pieces embedded in the actual install code. Valencia got to work putting the program back together and then saving it with a hidden name in an area that would not be likely to arouse suspicion. A quick edit of the batch scheduler would cause the program to be run four times a day. When running, a copy of any new files in the directories named would be copied offsite to a location where Valencia could retrieve them.

Only ten minutes had elapsed while getting that set up, and then Valencia unpaused the regular install. While it ran, he went exploring to see what else he might find. The terminal he was using had already been logged in, and when he checked, he had limited root permissions, which should allow him to poke his nose into just about anything. The first place he wanted to look was in the personal accounts of the senior board members. Amanda Forrest's files were locked up tight, as were Mathias'. Cynthia, however, had not been very thorough about encryption protocols. Within a minute, Valencia had a local copy of all her files – now what to do with it?

He couldn't forward it directly anywhere with her personal markers still part of the files. He couldn't strip the markers out without more tools than he had available, or more time than he had. He didn't have any physical media to transfer them to for smuggling out…Wait.

Valencia checked for progress on the install. Only 17% done. He killed that process and transferred the contents of the stick to a local drive and kicked it off from there. Then he erased the stick and went to move the files to it. Not enough room, Valencia decided, and he resorted the list to include everything with an access flag for the last six months. That fit, as did month seven and a fraction of eight. Once copied, he pulled the stick out and put it back in his sleeve pocket.

The actuator clicked, and "Erickson" was leaned back in his chair, fingers interlaced behind his head, when Jones came in. "How's it going?" the guard asked.

"Not bad. The manual stuff is all done, so the rest just needs to finish running."

"Sarge wants to know how long before you're done, so we can escort you out."

Valencia looked at the window with the install, with lines of file names flying upward. He switched to another view with a progress bar and said, "Looks like about five more minutes. You want to wait?"

"Nope. It's Latham's turn for that. I've got to finish rounds."

"Okay. See you next time, then."

Everything was done before Latham returned. He walked out with the man close by his right elbow. Unlike Verchou, Valencia could disarm and disable Latham before he would have a chance to react. Not that there was any need now, and hopefully wouldn't be. Latham walked him all the way back to the main elevators and watched him board. The doors shut, and Valencia was dropped down into unsecured territory.

He retraced his steps then, first to the service office to clock out, then back to the locker room to change out of his work uniform – transferring the datastick to his sock, and finally back through the security checkpoint of the government sector. There were no unanticipated events at any of them. Once outside the security gate, he checked the time and did some quick calculating. It was 1745, and the heaviest traffic would only last another half hour or so. He definitely wanted to be moving around in the bustle of shift change, but would it be better to make a direct run for the train station, or be slow and cautious and exit during the 0800 rush?

He finally decided to risk the speed and try for an immediate egress. There would be more passengers heading out of the complex in the evening, with most heading home from work, than there would be in the morning, when day laborers would be coming in.

With that plan in mind, he headed for the closest elevator, then waited in line with all the others still heading home. He blended in with the masses, moving along at the same pace and following the same routes. The elevators would only take him down two levels at a time, so he had to take five of them before he reached alpha level. Everyone in his elevator car went toward the station, and Valencia just tagged along. There was no security check at the outbound lanes, so his only delay was stopping at the machine to buy a ticket to the Walton complex in Chicago.

His train floated in, he boarded, the doors closed, and he was gone.

CHAPTER 34

12 October
Forrest Family Main Enclave, Earth

Amanda Forrest sat back in her favorite armchair and gazed out the penthouse window overlooking the darkened skyline of Raleigh. The walls of the Forrest enclave blocked the view of the century-old destruction to the south, but watching the lights of the slowly gliding traffic over the revitalized sections to the east and north always helped her relax. The Raleigh enclave was Forrest's oldest, built during the reconstruction following the Eleven Day War in 2366. As a major manufacturing concern, the Forrest Corporation had been one of the few entities with the financial resources available to begin any sort of rebuilding. As a consequence, Forrest had obtained even greater legal concessions than Warner or Sterling had been able to receive fifty years earlier when they had constructed the first non-government enclaves in Miami and Oxford. The remaining government was anxious for anyone to step in and establish some sort of order to the chaos that reigned at the end of the brief, violent conflict. As a result, Forrest had complete autonomy within their enclave and the government held no rights or authority there. They were not allowed even to enter the facility without permission. This had been a critical turning point in the growth of the Families and their eventual place as de facto rulers of the planet. The civil governments had ceded power to govern within the enclaves and the chaos outside those protected environments had continued until only the enclaves remained as viable entities.

Forrest now had twenty-seven operating enclaves around the globe, but Raleigh was the first and still the seat of power for the Forrest Family.

Amanda continued to watch the traffic as the minutes passed slowly by. She felt that the Families were at an equally critical point in their growth now. The four spacefaring Families dominated the worlds outside the Sol system and they would continue to grow and gain wealth as each world was able to develop and contribute. Forrest was not one of those Families that had claimed extra-solar property over the last century and a half. A good portion of their wealth came from shipbuilding and ship system engineering, but they had no official possessions beyond Earth. They ranked ninth in wealth among the seventeen Families, but that would continue to drop as the other Families took a greater and greater share of the pie with each passing year. The tariffs alone that Forrest was forced to pay just to transport goods through the jump points would guarantee their ruin eventually, as more and more of the population moved out to the stars. While this was not a problem yet, it certainly would be within the next century. This was why the Forrest Board of Directors had decided on their present course of action.

Amanda watched lights begin to go off throughout the enclave and in the city beyond as the sun cleared the horizon and decided that she could not postpone her next step any longer. While she had personally disagreed with the Board that this was the correct course, at the time she had been too new in her role as chairwoman to effectively oppose it. Now that the Board had made its decision, and things had progressed as far as they had, she had no options but to push events toward success for Forrest. With a deep sigh, she stood and walked back to her desk on the far side of the office. Forrest had taken the lead in the effort to even the playing field with Warner, Portales, Fermi and Sterling and it was time to begin the next phase of that effort, no matter how personally distasteful it was.

"James," she said as she activated the intercom in her implanted skull phone, "please hold all my calls and see that I am not disturbed for the next hour."

"Yes, ma'am," came the instant reply.

She sat down at her desk and activated the encrypted communications set built into its surface. She keyed the code for Felix Rial and waited while she was connected to the man who was not only the CEO of the Rial Family but also the head of the Families Ruling Council.

"Amanda," he said jovially as soon as the connection was established, "how are you today?"

"I am well, and you," she replied to the thin gaunt face on her screen that was at odds with the deep booming voice. "How is Heidi?"

"She is missing Gretchen now that she's finally off to University. Heidi still hasn't forgiven me for allowing her only daughter to go to Cairo for schooling instead of staying here in Geneva, but Gretchen wants to go into habitat design and Sterling is still the best source for that education," he said with a shrug. "And how is Marcus?"

"He's doing well," she replied. "Currently on Mars dealing with that labor issue. but he feels he should have everything wrapped up by the end of the week."

"Wonderful," Rial said, his eyes sharpening slightly. "Now, what can I do for you, Amanda?"

"Well, I was just wondering what you were going to do about this Warner issue."

Felix Rial's face froze for the slightest fraction of a second before he answered. Amanda knew he didn't like to acknowledge not knowing something. It was a small vanity but one that she could exploit.

"I haven't seen a full report yet," he temporized. "What were your thoughts?"

"Well, obviously I am very concerned that Warner is systematically harassing and damaging our shipping. From the way it is happening all across the frontier, this has to be some sort of cogent plan and not isolated incidents."

Felix Rial held very still. He was trying not to show his surprise at the accusations. "I had not heard that things had progressed so far," he said finally.

Of course, you haven't, she thought, *you haven't heard anything yet. I've just made it up.*

"It really is intolerable," she continued. "I have reports of several ships being delayed long enough to cause spoilage, others harassed into altercations, one ship that was impounded for unspecified violations. I think they are preparing to close their jump points completely."

"Surely things aren't that bad yet. Let me investigate and get to the bottom of this."

"We don't have time for long, drawn out investigations, Felix. I think we may be on the verge of another Rebellion. We need to take action now, before this gets too far out of hand."

"Let's not act without all of the facts, either," Felix said, his gaze sharpening at her continued push.

"Of course not, Felix," she said smoothly. "I think all sorts of investigations would be in order. Have you heard anything from Fermi, lately?"

His eyes narrowed to a slit and any trace of his earlier affability was erased from his features as her dart hit home. She was not sure what he actually had going on in the Fermi camp, but his quick reaction said it was serious.

"I need these sanctions ratified immediately, Felix," she continued, pressing her point before he could find a way to counter her push..

Felix Rial searched her features for any clue to how much she knew, how much she suspected, and how much she could prove. Finally, he dropped his gaze.

"I'll take it to the Council next week."

"And support the resolution strongly."

"And support it," he said as he switched off the comm without any further discussion.

She sighed as she shut off her own equipment. She had just turned a friendly associate into an enemy. If this plan did not come off successfully, she would have no friends left to turn to.

CHAPTER 35
17 October
Yargus

"I understand, Nigel, that in ground attack scenarios it is important to surround the enemy, or at least limit their avenues of movement and escape, but that is not always possible in a space battle; especially when we will be going in with only one ship. How do we surround someone with one ship?" Ensign Jherri Roberts asked with a grin on her face. Her ribbing of the Marine Lieutenant had become a nightly ritual after their sparring sessions. She had begun physically sparring with the tall, lithe Marine several weeks ago when she found his skill level to be near her own. The verbal sparring came as a direct result of the physical. On the mat, he was aggressive and inventive, just what you would expect from a Marine officer. Outside the sparring ring, however, he was reserved and quiet. She had begun the teasing as a way to get some response from the dour officer. Once he began to be more comfortable with her, his true personality emerged and she found him to be incredibly bright, drolly humorous and very insightful.

"Yes, I had heard that fleet officers were not capable of demanding tactics," Lieutenant Ndembe replied without any trace of a grin on his dark, African features, his upper-class British accent smooth and formal.

"Ouch."

"After you," he said as they stopped in front of the hatch to the conference room that would house the briefing for the coming operation.

The two junior officers entered the conference room and made their separate ways to their respective areas; Roberts to the left of the commodore and Ndembe to a seat near the wall, behind Major Chowdhury. The fleet officers were quiet and looking warily at the Marine major with the kind of sidelong glances you would use to keep track of a dangerous animal whose owner swore it was tame. It was almost as if they expected the major to draw her weapon and shoot them all without provocation.

Roberts understood some of their trepidation after the scathing commentary Chowdhury had given them all at their initial planning session, but she didn't understand their underlying attitude. Chowdhury was a consummate professional and had proven, repeatedly, her competence. After the shock of her statements in that previous conference, she had not repeated it in the week since. It had only been intended to force them to look beyond their preconceptions. How could they doubt her ability or her trustworthiness?

"Lieutenant Burkhalter," Brighton called, silencing the muted buzz of conversation, "can you finally sign off on the jump-engines?"

Lieutenant Shannara Burkhalter leaned back in her chair and ran her fingers through her long blonde hair as she let out a long slow breath. All eyes in the compartment turned to her as she pronounced the verdict on which rested all of their plans.

"Sir," she began tentatively, "all the tests show positive results. We have tracked down the fault in the capacitor system that gave us problems before and all indications are that the drive is up and ready."

Commodore Brighton nodded to the group and undoubtedly was thinking the same as most occupants in the room. All indications had been positive before their jump into Betre and into Antoc. Neither of those jumps had come off as planned, but this time there was no fallback position. If the jump engines didn't perform, they would not be able to follow *Pathfinder* and the pirates would escape.

Brighton nodded again and turned to Chowdhury. "Major, are you satisfied with the plans to reboard *Pathfinder* and take back the ship?"

"Yes, sir. If you can deal with whatever force the Forresters and DaGamans have in-system, we can take back *Pathfinder*," she said emphatically.

"Very well," he began, turning his attention to the room in general. "I had not anticipated that we would have a delay this lengthy before being able to jump to the Worth system," Brighton said ignoring the wince from Burkhalter, "but we must proceed without further delay in order to be in a position to support *Dagger* upon her arrival. Any further delay will make it likely that she will arrive before we do. As we have all of her embarked Marines, she will be unable to take back *Pathfinder* without our support."

"As all of you have been made aware, there are several possibilities that may face us upon our arrival. We have contingencies for each eventuality that we could foresee, but it is possible the situation may be different than any of our contingencies, so be ready to change plans on the fly," he scanned the room to make sure all had taken in the possibility before continuing. "Having said that, I believe we will face at least one, but probably not more than three, Forrest ships. They may be purpose-built warships, they may be merchant vessels, or they may be merchant vessels which have been refitted and armed.

The latter is more likely, but we must be prepared for any of the three, or a fourth, unforeseen, option. Our first priority will be to find *Pathfinder* among all the shipping and plot an intercept course. Major, your Marines will be ready to board the assault shuttles as soon as the jump is complete."

Brighton continued to outline the plans once again as Roberts leaned back and felt her stomach developing the queasy feeling she had not experienced since their last escape from Antoc. She glanced across the compartment to where Lieutenant Ndembe was taking notes and wondered if more of her friends were about to die.

✶ ✶ ✶ ✶ ✶

Commodore Brighton stood with his right hand resting on the back of the command chair that had been installed on the bridge of *Yargus* and faced the front viewscreen with no trace of his anxiety displayed on his features. His uneasiness was not for the possibility of action on the other side of the jump, but more for the possibility that the jump engines would again fail. They still had no successful test of the engines, but if they must test the engines, the Worth jump point was as good a target as any other, and better mapped than most. Time was of the essence. The consequences of failure here would be fatal for his Family in the long run.

"Lieutenant Burkhalter," Captain Ramirez called out formally, "charge capacitors on the jump engine."

"Capacitors charged and ready," she called back without any delay. "The ship is secure and ready for jump."

Brighton reviewed his plans again in his mind, forming and discarding scenarios in an effort to occupy his mind as the captain called, "Helm, jump the ship."

All focus on the bridge went to the pilot as the slight wave of nausea indicated that the ship had jumped somewhere. Everyone held their breath in anticipation of the news from Ensign Judd at his position at the scan console.

"Oh, crap!" was his unofficial notification that all had not turned out as expected.

CHAPTER 36
17 October
Dagger

Captain Fyonna Johnson sat back in her command chair and watched the monitor which displayed their position within the Antoc system. The atmosphere on the bridge was calm, at least on the surface, but there was a tenseness underlying everything that was done, every order given and obeyed. They had been pushing their engine capacity beyond any reasonable limits and she knew the crew was nervous about that. In fact, many of the crew were starting to wonder about her ability; if not her sanity. So she sat back and did her best to project a calm serenity to the bridge crew. She knew enough about how ships worked to know that her demeanor and any actions would be circulated around to the rest of the crew. She glanced down at the monitor once again and noticed that the ship was nearing the true edge of the Antoc system.

"Clearing heliopause of the system," Ensign Hayes called from the scan console.

"Thank you, Ensign."

She turned to her comm officer, " Davis, get me Larsen on *Foundation*."

"Yes, ma'am. Ready."

"Commander Larsen," she began, "we are coming in fast; we should arrive at the gate at approximately 21:44 standard. We would like to transit immediately upon our arrival. What is the status of the gate, sir?"

She waited several minutes for the message to travel the distance to the JP, where the construction ship, *Foundation*, and her crew were working to finish the gate generators that would allow them to leave the system and begin their transit back to Earth.

After a few more minutes, the reply came from *Foundation*. "Captain Johnson, we have been very lucky and the gate is completed ahead of schedule. Be advised that you and I owe my shifts leads a few rounds of drinks! We still need to send the test probes through at this point, but I don't foresee any issues. The gate control system is only about halfway complete, but we can operate the controls in manual. We should be able to put you through as soon as you get here."

"Thank you, sir. Please extend my thanks to your crew; all of them. They'll have no trouble from me collecting on those drinks. Johnson, out," she said and sent the response to the gate before turning her attention back to her comm officer. "Davis. Please get me *Avram*."

"Yes, ma'am. Ready to send."

She pressed the button on her console to begin recording a message. They were still too far from *Avram* to have a conversation without interminable lags as the messages moved at light speed between the two ships. *Avram* was still near the asteroid base on the opposite side of the system from the JP. "Lieutenant Carmichael, we are clearing the edges of the system now and should rendezvous with you at the gate in twelve hours. If you are not already underway, please expedite your departure. Also, confirm prisoner status and your readiness to transit. Upon my arrival, we will embark all Marines from the gate onto *Dagger* and transit immediately to the Betre system and begin our journey to Gateway. There are no other changes to your current orders from the Commodore. Johnson, out."

Several minutes later she received the response, "Roger, *Dagger*, we are en route to the gate now; we should arrive two hours prior to your arrival. All prisoners are secured and we are ready in every detail for the trip to Earth. *Avram* clear."

Johnson leaned back in her command chair once again, planning her next moves. While she didn't have the uncanny mathematical genius of the Commodore, she was a fairly good astrogator and a great pilot. The last several months had shown her where her skills were strongest and where they were weaker. She was always trying to improve any weaknesses and so she opened a fresh window on her astrogation console. She input the variables of their course to the gate and set the comp to working on the shortest time course. While that was running she set up the problem again and began to work it by hand. The computer would assume the engineering specs to be unassailable and would not propose any course that exceeded them. Her plan would assume a continued boost at 118% of rated power and a turnover and braking maneuver using that same power setting. It was a straightforward problem and she heard the quiet tone from the computer before she even finished setting up her own problem. She ignored the computer solution while she worked through her own problem and finally arrived at an answer. She was happy to see that her solution matched their current course and only differed from the computer recommendation by the amount of engine boost.

She put away her astrogation exercise and set to work on her other duties. They had turned over and begun to brake several days ago so their course was essentially unchangeable if they wanted to be slow enough to go through the gate when they got there.

"Comm, prepare a copy of all logs and load them in a buoy for the Commodore. Also, be sure to include our present schedule. Launch it when ready."

"Aye-aye, ma'am." Davis acknowledged.

* * * * *

"Comm, contact *Avram*," Johnson said as she re-entered the bridge several hours later. She glanced up at the chrono just as it hit 20:00.

"*Avram*, ma'am," Lieutenant Strachowitz, comm officer for this watch, answered.

"Captain Carmichael, please begin transferring Marines to this ship and make sure all available troops are retrieved from the gate control. Leave Larsen a squad to protect the gate and embark the remainder."

"Aye-aye, ma'am," Lieutenant Carmichael said in response. "We'll match our movements to *Dagger*."

As they approached the JP, they continued to slow and approached the point in space where they could detect the energy warping that indicated the Jump Point. As they slowed to enter the narrow gate, the new gate generator station energized for the first time in earnest and propelled *Avram* and then *Dagger* back into the Betre system that they had left 55 days before.

"Comm, contact Betre control."

"Ready, ma'am."

Betre Control, this is Captain Johnson in temporary command of Task Force Ten. Please clear JP3 for Military Priority traffic in sixteen hours. We will be transiting to Gregor en route to Gateway. Please relay this message up the chain. Authorization Zulu Tango Prime."

"Helm, give me maximum power to the engines with turnover in seven point nine hours."

"Aye-aye, ma'am."

With that taken care of, she leaned back and took a deep breath. The itinerary she had sent to Commodore Brighton had included two separate timetables. The first had assumed they could not get the jump engines working in time and would have to make the entire trip through the jump gates. The second had assumed that they would get the engines going at some point and would be able to jump directly to Worth. This second timetable was mere guesswork as she could not determine with any accuracy when or if they would be able to make the engines work. She knew that he would be counting on her to get there as quickly as possible to help reinforce them in any situation and she intended to do just that. However, regardless of the fact that they had created both timetables, she had always believed that they would be able to make the direct jump and cut a number of days off of the trip. She worried that *Yargus* would find herself in over her head and they would not be there to help. Now she had a report from Ensign Mitchell, endorsed by Warrant Long, that he didn't believe it possible to construct jump engines that would work on a ship as large as *Dagger*.

Lieutenant Reed had finally been convinced, and had added his endorsement as well. Fyonna had suspected there might be lingering bad feelings between the two officers and was about to ask Grant to keep an eye out for such, when her XO had surprised her with Reed's request that a note of commendation be attached to Mitchell's service record. When she had discussed it with Grant, he'd said only, "Tom will always hold a grudge against the young man but he respects solid engineering."

If *Yargus* had kept to her schedule, Commodore Brighton would have jumped into Worth six hours ago to an unknown reception. She had another fifteen and a half hours to JP3 in this system and a total of 56 hours to Worth at their highest speed and using the emergency priority.

She hoped she would be in time.

CHAPTER 37

18 October
Warner Naval Building – Quito, Earth

Admiral Cosina stood from his desk chair more slowly than he would have liked and returned a crisp salute from Col. Valencia as he entered and came to attention near the doorway. The admiral hated days like today when he felt his age. Once the salute had been acknowledged and returned, Valencia moved swiftly to stand in front of the Admiral's desk and waited as Cosina retook his seat and indicated the chair next to the colonel in a silent request to be seated.

No sooner had Valencia taken his seat than he handed the admiral a small data chip, and began explaining his findings. Admiral Cosina had always liked Valencia's ability to cut to the heart of the matter, with little need for pleasantries or even any questions from his superior.

"Sir, we have corroborating evidence that DaGama planted Jhonsruud in Warner service over nine years ago. This was an important part, but only a part, of their efforts to break into extrasolar operations. We have supporting evidence of other facets of those efforts as well," Valencia stated while handing over a second chip containing that evidence. A small nod from Cosina was correctly interpreted as permission to continue.

"The Forrest Family has been working very hard for the last twelve years to find any way around paying the tariffs imposed for transporting products through our gates and the gates of others of the Spacefaring Families. From the evidence here, we know they were able to infiltrate Fermi and Sterling as well. We have confirmation that they are also behind the theft of *Pathfinder* and the crimes committed during that action," the colonel continued in the same hurried but confident tone that was his norm.

At this point, the colonel cocked his head slightly to the right and back, seeming to point his beak-like nose at his commanding officer, and slowed his vocal pace as he went on.

"Sir, now we move more to the arena of conjecture without as much supportive evidence, but all indications lead us to conclude that Forrest, DaGama and possibly others have been in secret collusion to plan and execute a military takeover of at least one outlying system," Valencia stated and then paused as his CO leaned forward heavily onto the desk.

"How sure are we of this?" Cosina asked bluntly.

"As I said, I don't have enough evidence to prove beyond a reasonable doubt in a court of law, but there is sufficient evidence to convince me that they have planned this, and intend to carry it out. I posit that they will take it, hold and fortify it, and then announce their ownership hoping to get off by paying some small token reparations," his head of intelligence replied.

"You think they have grown so brazen? They previously have not moved with such open defiance and have always preferred the background game," Cosina queried him, but his tone suggested he also agreed with the assessment.

"Sir, I do believe it, and I have some evidence to confirm it. Not enough yet to take to a tribunal, but enough to convince me it is, in fact, their plan. The truth is, as we know, the Ruling Council is not likely to desire an escalation of conflict, and the conspirators are banking on the fact the others will acquiesce rather than allow outright hostilities to escalate. No one in their right mind wants open conflict between the Families.

"It's actually a fairly logical operation from their point of view. They are currently locked out of any way to expand their operations beyond Earth and current holdings, with small potential to make any inroads even on Earth. All of their trade goods are tariffed by us, and the other Spacefaring Families. Once they take and hold real estate out there, and once they can take care of what they perceive to be a bitter but resolvable period of reparation negotiations in the RC, they can make their move. They can begin expanding, and collecting their own tariffs to their systems, while paying less to the rest of us.

"Sir, to them, this is just business. Bitter, nasty business, but just business," he concluded.

Cosina leaned back in his chair, but only half turned toward the windows as he considered the import of the evaluation Valencia had just delivered. He finally harrumphed lightly and turned back to the head of intelligence and asked, "How much of this evidence can we use with the RC without compromising our sources?"

"All of the evidence against DaGama is verifiable from multiple sources, and we can keep it from tracing to anyone in a precarious position. Almost none of the rest of the intelligence on Forrest is safe to expose. I recognize some of it may have to be, but please weigh that carefully. Time sensitive as the gathering has been, we haven't found alternate sources for much of it," his intelligence lead left unspoken the *I shouldn't need to tell you how much our assets stand to lose if we expose them* that was evident in his intense, deep brown eyes.

Cosina accepted these statements and left unspoken his need for further corroboration. After a brief pause, he continued his questions for his spymaster.

"Which systems does our intelligence suggest may be targeted?"

"We're not yet sure sir," he stated flatly without making any apology for his lack of knowledge. His competence was a known quantity, and he never felt the need to apologize for not having all the needed info. What was, simply was. The colonel waited quietly while Cosina chewed on this.

"I would add sir, that there is actually a significant likelihood that they may have already struck such a system, and we are just blind to where it has happened. There are far too many involved parties, and plots like this have a way of coming unraveled at the edges, only to take on slightly different thread patterns than anyone anticipated."

"I am thinking much the same thing. Colonel, use whatever resources you can safely and securely involve and dig more deeply into that aspect. Keep it as quiet as possible, but again, the information and the speed of its availability are more important to me right now. Keep me apprised of any significant details," Cosina said, clearly indicating the briefing was over. "Yes, Sir," Col. Stefan Valencia rose swiftly and snapped a salute as his CO was still coming out of his chair. It was returned by his boss, and Cosina told him he was dismissed.

Valencia lithely spun about and moved from the admiral's office with purpose.

Before he was completely back in his chair, the admiral had pressed his comm into service and requested that his exec get his Director of Development & Engineering on the line immediately.

About thirty-five seconds later his comm chirped and he heard the familiar voice come across it asking what was needed.

"Operation Caliban is a go. I want a revised timeline in my hands in thirty minutes, and I want you in my office in two hours to walk through the details, with an eye to cut out every wasted moment in the plan," the Admiral stated, his voice gruff and uncompromising.

"Order acknowledged sir," his DDE said after a slight pause. "Caliban is a go. Your timeline brief report in 30 minutes, and I will be there in two hours to discuss in person. DDE out."

Admiral Cosina once again rose from his chair, and turned to pace slowly along his window wall. He ran through possible scenarios in his mind. While looking out the windows, he didn't really see anything. His thoughts were far too focused and his brain was ignoring irrelevant sensory details. Operation Caliban would give them four more jump capable warships. If the previously assessed project timelines were held, they might be done in time. He had a twisting in the pit of his stomach that told him he'd need them very soon. Maybe sooner than he imagined they would. That timeline was going to have to be shaved, and while no one would like it, they'd have to find a way to get it done.

He wasn't one for melodrama, but in fact, the entirety of their future as one of the most powerful and influential members of the RC, and what that meant for their millions of citizens, was in jeopardy if he failed them. Not to mention, more importantly, the thousands of lives that might be cut short as well. His team would deliver. They had no choice.

CHAPTER 38
Forrest Headquarters
18 October

Amanda Forrest looked up from her desk as her assistant, James Wight, ushered Mathias Kinsey-Forrest into her office. Mat was currently serving as the Vice-President for Off-Planet Affairs. As such, he was the person responsible for much of the current activity to offset Warner advantages in space. He was short and slim, wearing a perfectly tailored, very expensive suit. His thinning hair was slicked straight back into a two inch queue at the nape of his neck.

Mat had commed her office asking for this meeting only two hours ago, so there must have been some change in the status of their current plans.

Amanda was still uneasy about the lack of any hard data on the status of *Pathfinder*. She had monitored the media very closely, but so far Forrest had not been linked to any of the stories. The return of William Brighton and his crew had caused a stir back in August, as had his subsequent departure with the majority of Warner's available fleet. There had been no further reports on his whereabouts or actions, and that made her nervous.

She rose from her desk and motioned Mat over to her reception area. When they were comfortably seated, she waited for his report.

"*Pathfinder* finally showed up," he said in the clipped, nearly angry tones that were his trademark.

"In Worth?"

"Yes, it jumped in on 6 October."

"Well, that's certainly good news."

"Yes and no," he quipped with distaste.

"Why not?" she asked, her eyes narrowing in suspicion. Amanda knew that she viewed everything that came from Kinsey-Forrest with a large degree of suspicion and skepticism. It was no secret that Mathias felt he should be CEO. He was an expert in the political in-fighting common to any boardroom. Anything he could find to discredit Amanda he would use without hesitation.

"The ship arrived, finally, but it is not in good shape."

"It doesn't matter. We have the ship, we have the plans and we have a way out from under the Warner monopoly."

"We have the ship," he corrected, "but we don't have the plans. The ship is a mess and it will take us months to salvage anything out of her."

"If the ship was able to jump into Worth, then the engines must be intact and functioning. If the engines are intact, we will be able to reverse engineer the systems. We have the facilities of a complete shipyard at our disposal as well as the assets of an entire system. Unless, for some reason, you cannot deliver on your promises to the Board."

The jab was well-aimed, and Mat knew it. it had been he who had first gotten wind of the Argo project, and had pitched the idea to the board of stealing the new technology. Any complications, therefore, would serve to undermine his own position. Kinsey-Forrest leaned back into the plush sofa, as if he were aware of none of that, and looked into her eyes. His expression was tense and his posture was aggressive, despite his apparent ease.

"I just received the latest word from your girl Solomon on the planet," he said, moving his briefing folder slightly on his lap.

Amanda caught the emphasis on 'your girl'. Whenever it was bad news, Epi Solomon was 'her girl,' but whenever there were significant accomplishments, she was 'his associate.'

"And what have you heard?"

" The resistance on the planet is increasing. They have had numerous disruptions at the shipyards and the schedule has slipped again. They won't have the first ships retrofitted for several more weeks, if then. She has requested more troops."

Mat paused for a long time, weighing whether to go ahead with this gamble, now that the moment was here. "She also admits that she is barely adequate to the task and has resigned her authority, turning it over to the military commanders on Granada."

Amanda was silent for several moments as she absorbed the information. This was disturbing news indeed, and she would not have believed Solomon would ever back down from a challenge. The timing of her resignation could not have been worse, either. They were quickly getting to the point where their actions would come to light. Nothing done on the scale that they were working could be kept quiet indefinitely. They had invaded systems owned by several other Families. Someone would trip over their presence eventually, regardless of how carefully they guarded the secret. They needed to attack and cripple the fleets of the other families before they could unite against them. This had always been a part of the plan.

In order for them to be able to do this, they needed ships, many more ships than they currently had available. She did not indicate how troublesome this news was at this point in the plan. After years on the Board, her face gave away no trace of what was going on in her mind.

She quickly came to a decision. Warner was getting close, but she wasn't quite ready to openly challenge them yet. She had a contingency plan that should keep them off balance for the time necessary to finish their preparations. In the meantime, they needed Worth under control.

"Which specific officers did she leave in control?" the CEO asked.

"Admiral Franks is the ranking naval officer, and General Franks is in command of the Marines."

Franks? Amanda thought. Having met the woman, she could imagine that Solomon and Franks would get along like a cobra and a mongoose. If the situation were bad enough for Epi to admit defeat to... No, she never would. Something was wrong with the picture Kinsey-Forrest was painting, and Amanda would get to the bottom of it. She filed that away to work on in private and brought herself back to immediate concerns.

"You are authorized to send an additional eight thousand troops to Worth, but you tell those two officers that we simply cannot afford any further delays. They absolutely need to produce the required ships on their due date or I will have them shot."

"Are you sure that's wise?" Mat asked with wide eyes. "They know best what is going on there and what it will be possible to deliver."

"Mat," she said, stabbing him with her icy glare, "we are out of 'wise' choices. We have no choice at this point. I'll not take pity on anyone who can't deliver on their promises." It was clear that she referred to more than just Solomon.

Mat Kinsey-Forrest rose with a slight nod to her and moved toward the door. It opened just before he reached for the control and James stood holding it open until he was gone.

Wight entered, secured the door and moved to the end of the sofa where Amanda sat. He stood quietly and watched her as she stared out the window. Her right leg was slowly swinging from where she had it crossed over her left knee. The fingers on her left hand were slowly tapping on the arm of the sofa. Her assistant was sure that she was not aware of either habit, but he knew them for what they were. These were mechanisms that she used when she was marshalling her thoughts and tracking possibilities. He had been with her long enough to know that this was her way of putting all her ducks in a row, as the old saying went. She would let him know what needed to be done when she was ready.

"James," she said finally.

"Yes, ma'am."

"Send a message to Montague, Cadat, and Knox. Tell them to escalate their building pace. Also tell them to increase security and ensure that no one outside their systems get suspicious. Make sure you send that with the highest urgency.

"Also, send word to Stavros to contact me. I have a suspicion that Mr. Kinsey-Forrest may be making his first moves in a takeover attempt."

"Yes, ma'am. Anything else?"

"Yes, send a message to Director Rial and formally protest the uncivilized treatment of the crew of *Hermes* by the Warners at Madsen. When those tasks are completed please let chef know that I will be taking dinner at my desk again this evening."

"Yes, ma'am," he said as his boss rose from her sofa and moved back to her desk. She never even noticed as he let himself out to complete her assignments.

CHAPTER 39

18 October

Yargus

"Oh, crap!"

The exclamation filled the silent bridge as the scans cleared and the nine unidentified destroyer-sized icons lit up on the screen.

Commodore William Brighton shook his head to clear it of the effects of jumping into the Worth system and decided not to chastise the embarrassed ensign at the scan console. The officer in question had turned a bright shade of crimson as all eyes on the bridge turned to gaze at him, and the annoyance on the face of Captain Ramirez let the commodore know that the captain would have words with the young man at the earliest opportunity. That didn't change the fact that his choice of words seemed to sum up the change of status for the undersized destroyer they were in.

"Commodore. Message coming in from one of the ships," Ensign Lyle said from his position on the comm console. His thin, pale face seemed to have gone another shade lighter as he hit the control to play the message at the commodore's motion to do so.

"Unidentified ship, this is FFF *Hammer*. Cease all movement and prepare to be boarded," came the stern voice from the speaker.

The ensign turned to the commodore as if waiting for a response. Brighton made no move for several moments then stood and began pacing behind his chair. Finally, he stopped and signaled to the ensign and, at his nod, said, "Unknown ship *Hammer*, this is Commodore William Brighton, commanding Task Force Ten of the Warner Space Navy. Stand down your weapons and prepare to give an explanation of your actions in this system."

It was a stalling tactic. One meant to turn the tables on the invader and make them pause long enough for Brighton to decide on the proper response and the best way to achieve his ultimate goal of recovering *Pathfinder*.

His mind flew through all his prepared scenarios within a few moments as he considered this new development, what it meant to his mission, and how to fit it into what responses they had set in place before the jump.

Even though he and Major Chowdhury had pushed the planning staff to consider such unlikely events, no one had foreseen *this* many armed combatants being in the Worth System. He had been most concerned about the jump engines on *Yargus* and had thought that, once here, the mission would be fairly straightforward.

Despite the limited testing of the jump engines installed in *Yargus*, they seemed to have performed perfectly this time. They had entered the Worth system right on the jump point and had transitioned into normal space without incident.

It would appear as if that were the extent of *Yargus'* good luck, however. Instead of a Warner system able to help them with their chase, they had dropped into a system that had been taken over by a foreign power. There were at least nine warships active in the system with several more near the planet that were too distant to be identified yet. In addition, the warships had proven themselves to be hostile by issuing a challenge to his ship immediately, without waiting for any kind of identification. Such actions could only mean that all Warner defenses in this system must have been disabled or destroyed.

He thought again about that demand. FFF *Hammer*? Was that a Forrester designation? There had been evidence in Antoc that the Forrest Family had been involved in the operation at some level, but the majority of the activity there had seemed to have been perpetrated by the DaGama Family. He had thought them to be behind the theft of *Pathfinder*.

Brighton briefly considered his responses. He was confident in this ship and its crew, but they were still unproven in any kind of battle. Another factor was the level of forces on each side. His scans of the system were starting to clarify on the repeater screen to the left of his seat and he quickly identified *Pathfinder* in an orbit very loosely circling Granada, the only habitable planet in the system. The hostile warships had been patrolling in a position to cut off any ships entering the system from both the planet and from returning to the gate and jumping back out of the system to get help. That left few options to most ships transiting into the system and he made a quick count of eight freighters that seemed to be in permanent orbit around the planet and guessed that their captains had been forced to surrender to the invaders. He had many more options than they had had, however.

"Captain," he began, looking at Captain Ramirez, "please change course toward the ship marked on your scan as bogey eleven. Also, change the designation of that ship to be identified as WNS *Pathfinder*. Watch for any changes in the formation of the Forrester ships and stay out of their weapons envelope."

"Sir," the captain began, "we are heavily outnumbered, wouldn't the prudent course be to head back to the JP and get word to Fleet HQ?" He struggled to keep his voice even, but the strain and worry in his stance spoke of his nervousness. As much as *Yargus'* crew were unproven, the same could be said for her captain. Having taken an engineering track to his current position, his command experience was very limited, and did not include any type of hostile encounter.

"That is part of our responsibility, yes," Brighton began, "but I cannot overstress the importance of recovering *Pathfinder*. If we have to sacrifice this ship to retrieve that one from the Forresters, it would be a favorable trade."

"Sir, we cannot take on that many ships and we cannot evade them indefinitely."

"No, but we can assess their level of commitment."

"Sir, I–" he began before being cut off by Brighton.

"Captain. You have received your orders," Brighton said in a quiet voice that was deadly in its stillness.

"Aye-aye, sir," he replied and began making the necessary orders to his crew.

As the ship began adjusting its course, the Forrester ships moved quickly to cut them off.

Brighton nodded and said, "Captain, make your course 108. 223.00, on the system elliptic. Contact Major Chowdhury and tell her to prepare her Marines to board their shuttles and then charge your capacitors."

"Aye, sir," Ramirez said tentatively, as if there were something there that he should understand but was not grasping.

Brighton returned to his seat and nodded softly to himself. *Yes, that should work*, he thought.

CHAPTER 40
18 October
Yargus

Commodore William Brighton ignored the bewilderment that was clearly evident on every face around him on the bridge of *Yargus*. He sat perfectly still in the command chair that had been installed on the starboard side of the bridge. He glanced to his right at the screen that duplicated those found next to the captain's own chair. He nodded slightly, as if confirming the decisions that he was about to implement.

"Captain Ramirez," he began "make the ship ready for combat."

"Commodore," the captain said gravely, "I don't know what you're thinking, but taking on nine destroyers is suicide. We cannot possibly defend ourselves against so many of the enemy. We must retire while we can and get word back to Fleet Headquarters. Please reconsider your orders and do not let your obsession doom this ship. If we die here, we will have accomplished exactly nothing."

Silence reigned on the bridge as each officer and crewman sat tensely quiet at their station while their superiors decided their fate.

"Captain, I am not in the habit of repeating myself," the commodore said as he struggled with his temper. "You will execute my orders this instant, or you will find yourself in violation of Article 23 of the Fleet Regulations. As I now consider this ship to be in a state of war, the penalties pertaining to that Article are extremely stiff. Captain... for the last time, make this ship ready for combat!"

It was quite possibly more than he should have said on the bridge of the ship where the captain was expected to command, but the only other possibility was to relieve the man, and under the circumstances, the young officer's career would not survive that event. It was possible, if Ramirez began acting like the capable officer he was, that no ill effects would come his way, but Brighton would not put *Yargus* at risk to try to protect a man who couldn't follow orders. Surely, the young captain could see that, while his only option was to submit and obey, Brighton's ability to overlook his outburst was similarly limited.

"Aye, aye, sir," came the stiff reply.

"Maintain a course directly away from the jump point at two-forty-three gravities," the commodore said calmly and courteously, as if the previous reprimand had never occurred.

"Aye, sir," the captain said before turning to the helm to relay the orders. "Charge all energy weapons. Load all missile tubes and set Condition One throughout the ship," the captain said strongly and clearly, with only a slight indication of the emotion he was struggling to restrain.

"Aye, aye, sir," came the chorus from the bridge crew as they studiously performed their appointed tasks.

"The two nearest destroyers are starting to close, sir," the young scan tech, Ensign Judd, said to his captain. "Designated Tango-1 and Tango-2. They must have been underway and should have a slight velocity advantage over the others."

"Are all the ships in pursuit? Did any remain near *Pathfinder*?" the commodore asked of his flag captain. He didn't want to undercut the man's authority any further by asking the questions directly.

"All are shaping their course to follow us, sir," Judd replied, also directing his response to his captain.

"None are remaining in the vicinity of *Pathfinder*, sir," Captain Ramirez relayed to Brighton who nodded absently as he studied his scan.

"Captain, fire two missiles at each of the two leading ships as soon as they come within maximum range, then secure your launchers."

Brighton continued to watch his scan and the various displays at his right as Captain Ramirez relayed the proper orders. Brighton's command chair, a difficult addition to a bridge never meant to hold a flag deck, sat above and to the right of the captain as well as slightly behind him on the outer ring of the bridge. From this vantage point, he could see all the other members of the bridge crew except the tactical officer, Lieutenant Kent Weaver, who was on the far side of the captain and blocked from his view. Those that he could see appeared to be comforted by the routine of their captain's orders. The content of those orders was anything but commonplace, however. Brighton doubted whether any of these officers or crew had ever been in anything other than a simulated battle, but they were comforted that their captain was giving orders and there was no forthcoming confrontation between a captain that they were just starting to know and a commodore who they knew only through reputation. They were settling into their jobs, but they were still slightly on edge.

For a moment, Brighton thought how different something was, and then realized that it was himself. In the past, the emotional condition of the bridge would have been the last thing he would have considered when heading into conflict. Weapons readiness, tactical positioning, even the crew's last performance rating were all things he would add into his calculations, but how the crew was *feeling*? This was the first indication he had noticed that his ordeal in the Antoc system had changed him.

"The two leading ships are coming into range now, sir," Tactical called from the far side of Ramirez.

"Fire missiles as planned, Lieutenant," the captain responded, while fighting the urge to glance at his superior.

"Missiles away, sir. All running straight and true. Forty-three seconds to impact."

Brighton looked around the bridge again. It was much smaller than the bridge of *Dagger*, which he had just been getting used to. Though *Yargus* was new to him in many ways, since its refit, it was still infinitely more familiar than *Dagger* had been. *Yargus* was of the same design as *Pathfinder* had been before her refit, the long-serving *Risea* class. The bridge configuration was much the same as *Pathfinder*, and still nearly the same (minus a flag seat) as she had been when she was Brighton's third command, *Reigna*, which he had commanded during that Humboldt fiasco so long ago As always, thoughts of that Humboldt mess brought a frown to his face, and he quickly jumped back to his earlier train of thought.

The captain's chair in *Yargus* was located on the port side of center in the middle of the bridge, where *Pathfinder* had located the command seat toward the starboard side of center to accommodate a larger, more advanced helm control which had replaced the old helm and tactical boards. There were several other less obvious changes, but the overall effect was one of familiarity. All the officers and crew manning the stations were different from the familiar complement of *Pathfinder* but, so far, they seemed to be equally competent. He was pleased to see that their settled emotions were allowing that competence to show.

"Two hits, sir. It doesn't look like serious damage. They're still coming, but they've slowed."

"Very well."

"Captain," Brighton called quietly, motioning the diminutive commanding officer over to his chair. "It appears that we are holding our own in acceleration. Hold a course directly away from them. If we turn, we will allow them to cut inside our turn and make up distance. We want to pull them as far from *Pathfinder* as possible before we make our move."

Brighton sat back and contemplated his plans. He had not shared those plans with his captain because he was still refining the data. He knew this was not fair to his flag captain, who should be able to direct the fighting of his ship. He was not able to do this without the knowledge that only Brighton possessed. Perhaps, if he had shared those plans he hadn't finalized yet with

Ramirez, that earlier confrontation could have been avoided. Still, while it wasn't fair to the captain, it was not an excuse for questioning orders, especially in front of their subordinates.

"The destroyers are closing the gap slightly, Captain," Judd called across the bridge. "They should be able to range on us in about four minutes."

"Helm, hold your course," the captain said firmly, as if he had never had any doubts about the plan they were following. "Engineering, I need every ounce of power you can muster. Go to full military power."

"Aye, aye, sir," replied the dark-skinned engineering tech.

"Targets one and two are still closing. Their acceleration is slightly higher than our own. The other seven are spreading out to cover all our possible courses back to the gate," Judd said.

"Continue to monitor their speed. Let me know when they reach maximum missile range again," the captain responded.

"Missiles inbound!" called Judd, his voice rising slightly.

"Helm, evasive plan alpha. Mr. Weaver, defensive weapons free. Fire at your discretion," called Ramirez.

"Aye, sir," they both replied in unison.

"Captain, you may return fire at your pleasure with the broadside weapons," said Brighton calmly.

"Aye, aye, sir."

"Helm, port 90 degrees. Lieutenant Weaver, fire your port broadside as it bears. Concentrate your fire on targets one, two and three."

Captain Ramirez settled into his command chair as he felt the six distinct launches of his port broadside. Lieutenant Weaver had expedited his launch to the minimum stagger between tubes. He was obviously aware how tenuous their particle shielding was at this velocity with the ship turned 90 degrees to their momentum.

"Missiles away. All running true. Thirty-five seconds to first impact."

"Clean miss on all eight inbounds. More launched. ETA of fifty seconds," Scan reported.

"Helm, come back to our base course. Lieutenant Weaver, continue your engagement with aft tubes. Minimum rate of fire. Let's conserve our munitions."

Brighton nodded to himself as he watched Captain Ramirez settle into his role. He seemed to have lost any trace of his earlier misgivings. The ship began to swing around slowly as the engines pushed the ship around to their previous course. *Yargus* had just begun her previous evasions when Brighton felt four jolts as the aft tubes began a staggered launch. They were followed by four more after several minutes pause.

"Several hits... it looks like two on target one and one on target three," called the tactical officer.

"Blast, one's through…" called PO2 Calvin Tomita from his console at laser control where he had been working to take out the incoming missiles.

Yargus shook as if her hull had been slammed by the hammer of some vengeful titan.

Brighton watched as Ramirez' face began to show some of its earlier concern. Ramirez glanced around tentatively, as if waiting for the follow-up blow that would erase them from the galaxy. Brighton knew from experience that *Yargus* could take more punishment than Ramirez seemed to believe. More even than Brighton had once believed possible. The *Risea*-class ships were tough. *Reigna, the ship that Yargus had started out as,* had saved the lives of many of his friends with its resiliency.

Two more missiles impacted in quick succession. Ramirez' face began to look ashen.

"Captain," Brighton called to focus the young officer's attention, "what is the status of your jump engines?"

"The jump engines are fine, no damage," the captain replied distractedly after consulting his screens. Brighton's own screens displayed the same information, of course. The captain resumed his position for a moment then suddenly sat bolt upright and turned to look over his shoulder at the commodore. A slow smile spread across his face. Brighton sat back and began to relax. Captain Ramirez had finally figured out his plan.

CHAPTER 41
18 October
Yargus

Commodore Brighton surreptitiously glanced at the chronometer mounted on the forward bulkhead. Nearly time, he decided. The thought was punctuated by another hit to *Yargus*. This one was much more severe and he struggled to maintain his seat in the command chair.

"The hostiles are spreading apart. It looks like they're trying to cut off our path back to the gate, sir," called PO2 Dan Jordan, Ensign Jud's replacement for this watch, through the scrambling of crewmen to regain their seats. He clung tightly to his console. Bridge chairs had always contained straps to help their occupants stay in place during violent maneuvers and especially during combat, but they were never used and most had been lost in the crevices of the padding. Crewmen were sliding hands along edges and behind cushions trying to locate these former nuisances which had suddenly become vital.

"Captain, change course back to the gate for ten seconds and then resume course. Let's try to isolate tangos one and two and get a shot at them," Brighton said calmly to his flag captain as he studied his scan console.

"Aye, sir," Ramirez responded as he located and secured his own belt. "Helm, come about to heading 124.22 relative, same plane. Maintain acceleration. Weapons, starboard broadside on targets one and two and fire as you acquire."

"Aye, aye," came the chorus.

Yargus turned cross-wise to her path of travel and a bit beyond, aiming her prow at the far distant jump point. As soon as the missile tubes were at an angle to engage, they began firing. Captain Ramirez waited until he felt the six distinct jolts that indicated all the broadside weapons had launched.

"Return to previous course," he ordered, not waiting for a report from his weapons officer.

"Aye-aye, sir," replied WO3 Reko M'buke from the helm.

"All starboard tubes launched, sir," called Lieutenant Weaver from his weapons console, confirming what the captain had already felt. "Forty-seven seconds to first impact."

"Captain," Brighton called, "Feel free to continue engagement with your stern tubes at your discretion. Let's try to reserve about twenty missiles for our stern tubes, if that is possible, however."

"Aye, sir," the captain replied looking over his shoulder at his superior. "Weaver, begin firing with stern tubes at two-minute intervals. Concentrate fire on target one."

"Aye-aye, sir."

"Incoming missiles. Launches detected from all ships. Plot shows thirty-one missiles inbound."

Brighton checked the chrono at the front of the bridge once again. These Forrest ships had demonstrated a slightly higher acceleration and top speed than he had anticipated, so he had to take them a little further from *Pathfinder* than he had originally planned. Of course, that would mean they would need to brake more heavily than he had planned as well. He spent a few seconds calculating the pursuit force's return vector and timing, and once again wondered if he were allowing sufficient time to complete what they needed to do before they could return. More Forrest missiles arrived, and defensive lasers lanced out to intercept them. The Forresters were as unused to combat, as were most of *Yargus'* complement, but they seemed to be adapting quickly. They were learning to send their missiles grouped in waves now, which made it more difficult for the defensive systems to stop them all. Inevitably, some were getting through with every wave. Brighton hung onto his seat arms as the bridge was rocked with explosions from the dual survivors of the salvo. Dust and debris rained down from every possible crevice and the environmental pumps could not keep up with the detritus. Dust motes sparkled in the light from the screens at his side as he began his calculations again.

"Damage control reports that missile three is out of action. Three casualties, no fatalities," WO3 Estella Gomez called from Engineering.

Captain Julio Ramirez relaxed his arms by conscious effort. He knew the crew would be watching him. He had already shamed himself with his earlier outburst and he vowed there would be no repeats of that disgrace.

Finally, Ramirez unfastened his restraints and walked slowly to Brighton's position. "Have we pulled them far enough out of position yet, sir? I'd like to get *Yargus* out of here before they get lucky," he said in a voice that was as respectful as he could make it. He did not want to be accused of insubordination and he had not figured out how to read the commodore yet. He seemed to be a cold fish most of the time, but there was obviously a controlled temper under the calm façade. *I need to figure out what the triggers are on that temper if I want to have a chance of being effective*, he thought.

215

"I understand your outlook, captain, but we are not quite into position yet. We need to hold this course for a few more minutes, at least, and that would provide us with no margin for error or Murphy."

"Should I have the helm plot a course for home?"

Commodore Brighton looked squarely into the face of his flag captain. It seemed as if the young officer's mood was completely changed from the tentative, panicked officer who had entered the system with him a short six hours ago. The actuality of battle had steadied the man; and his crew was taking their cue from him.

"No, captain," Brighton began, "since the Forrest Family is being so accommodating as to send every mobile warship after us, I see no reason not to jump back into the inner system and take our property away from them while they flounder about helplessly out here." Brighton smiled at the younger man, but there was no warmth in it; only cold determination. "We will not leave this system without *Pathfinder*. Your ship is performing marvelously and we should have no problems reaching our objectives. Now, let's try to take out one of those destroyers, if we can. With Forrest building warships, this is clearly a larger problem than we had envisioned at the outset, and any information we can gather on the enemy's performance and capabilities will be useful in the coming months. Get with your tactical officer and come up with a plan to separate out target one or two and destroy them without bringing excessive damage to *Yargus*."

"Aye-aye, sir," he replied as he moved off to complete his task. He suspected that the commodore had created the tactical problem simply to occupy his mind, but he was grateful for the distraction in any event.

Lieutenant Weaver, it turned out, tended to attack any problem as if it were a physical enemy that must be worked into submission. He was constantly making arm gestures and punctuating his points with pointed fingers as they worked out their strategy. Finally, after several false starts, they had a workable plan.

The time had continued to pass with an occasional launch from each group, while the officers had worked to develop their response. No further hits had been scored by either side. With both sides concentrating on the chase, neither was willing to lose ground by turning to fire broadsides, and the lower number of munitions that could be fired by stern or chase tubes was not enough to slip past very able defenses. Besides which, Ramirez had given orders previously to conserve *Yargus'* limited munitions, so their fire was meant only to make their pursuers as cautious as possible and to slow down their pursuit. *Yargus* only carried 82 missiles in her aft bunkers and she had already gone through nearly a third of that number. While it was possible to move missiles from the broadside bunkers to the aft stowage, it was a slow, tedious, one-missile-at-a-time process that was extremely labor intensive. With the crew at battle stations, the only bodies available to make the transfer were the damage control party, who were already engaged in dealing with the aftermath of the missile strikes already inflicted by the Forresters, and the Marines, who were on standby to repel boarders or assist in any of the weapons rooms where they might be needed, and had also been directed to be prepared to load up onto assault shuttles. Brighton had been adamant that the Marines were essential to his plan at a later point and he would not use them up now. So the missiles would stay in their current bunkers and they would just have to conserve them as much as possible.

When he finally presented his plan to Brighton, the commodore saw flaws immediately. "This plan is not without its risks, Captain," he said finally, "I think that you are counting rather heavily on their inexperience. However, I believe that it still has an excellent chance of success. Is there anything that you would like to add?"

"Yes, sir. I would like to target bogey four instead of the ships that you designated."

Brighton was nodding his head slowly as he asked, "Why would you make that change?"

"Well, look at the distance between four and six, and even more from four to five. They've spread out far enough that neither of the two nearest ships can assist in missile defense. Plus, if we are successful in destroying or crippling that ship, it would allow us to maneuver around their flank. If we combine that fact with a course change to 024.2 by 67, then we can put some extra distance between us and the main fleet. Those on the far flank will not even be able to range us until they can work their way around the main body. If we don't completely destroy four, however, we would be giving them a better shot at us."

"Can your crew successfully implement this plan?"

"Yes, sir," he replied emphatically.

"Then, please proceed, Captain."

"Aye-aye, sir," the captain said before turning to his engineering officer and beginning to implement their plan. "Gomez, I am going to need you to increase the power to our particle shielding during this evolution along the lower starboard quarter. That's going to put a strain on the engines, so keep a close watch on the numbers. Advise me before anything gets critical."

"Aye, sir."

"Helm, prepare to come starboard to 024.2 by 67. Hold us on that vector for thirty seconds at maximum, then align the ship perpendicular to the advancing enemy and give us a ten degree per second spin. Drop acceleration to 20 percent. Weaver, implement the fire plan as the target bears for each broadside. Thirty six seconds for a reload is going to mean those gun crews have to be flawless, but they should only need to do it twice."

Weaver looked at his captain in surprise for an instant, but realized what sort of confidence Ramirez was showing in the crew; confidence the lieutenant was not going to disappoint. "Aye, sir. Give me just a minute to set things up." He ducked his head over his comm link to the weapons rooms and began explaining what was going to be expected of all of them. A few seconds later he raised his head and nodded to the captain.

Ramirez paused and took a deep breath. "Now, Mr. M'buke."

The ship began moving to her new course and then the acceleration died away. The officers on the bridge felt nothing, but their displays told the full story of what was going on. The Gravitas plates in the hull maintained a constant gravitational pull regardless of the orientation or acceleration of the ship.

As her attitude changed and her gun ports were aligned, the missile tubes fired in one, ship-pounding, simultaneous launch. The target was the destroyer on the outer edge of the formation. Thus far, they had concentrated their fire on the inner three or four ships. The target should come as a surprise and, hopefully, confuse them and make it harder to coordinate their defense. They intended to make it even harder.

"Rotate the ship, Helm. Fire as the target is acquired, Lieutenant Weaver."

"Aye-aye, sir," came the responses, almost as one. The ship began a maneuver that would turn the ship over on its long axis. This would allow them to fire the port broadside at the target as well as the starboard, which currently faced the unlucky destroyer. The ship moved at the designated rotational speed in a way the crew had practiced hundreds of times in simulations and drills. The scans struggled to hold reference locks, but with the aid of their human operators, they managed. The platform stabilized for the briefest of moments and the port broadside was released as one. The ship was shaken again as the nine missiles were thrown clear.

Then the roll continued, bringing the starboard side back around, and nine more missiles headed toward tango-4. A roll to port, and nine more joined them. On the final spin, the pace turned out to be too much. The starboard crews only managed to launch five weapons, and the port side managed six.

"Humph. We'll need to work on that," Brighton commented, and Weaver's light complexion turned a deep red.

Redemption wasn't long in coming, however; travel time out to the edge of the formation was only a little over one minute. On his own initiative, Weaver had programmed the outgoing birds with different thruster instructions for each volley, so that the entire flight arrived at the designated target within seconds of each other. Forty-seven independently maneuvering missiles were simply more than a single destroyer could handle on its own, and eleven of them made it through to detonate against its hull. A series of flashes which turned battle armor into so much expanding plasma, and tango-4 was no more.

"Excellent work, Guns," Brighton said now. "Very commendable." Weaver beamed at the rare praise.

"Shielding status," the captain called to Gomez at Engineering.

"Holding, sir. There are some slight fluctuations, but nothing major."

"Helm, let's stabilize the ship and resume the last given heading."

"Aye, sir."

"Resume full military power as soon as you have the ship under control, Mr. M'buke."

Yargus began to pull away at a tangent, making the best use of the hole they had just punched in the net meant to contain them.

CHAPTER 42
18 October
Yargus

Commodore William J. Brighton fought to project an air of calm amid the chaos of the bridge. None of the crewmembers of *Yargus* had any previous experience with combat, so they would take their cues from him. He looked around the bridge at the officers and crew. They had been in a running battle, sporadic as it was, for nearly four hours. *Yargus* continued to race for the outer reaches of the system with the entire remaining complement of Forrest ships giving chase. Brighton had successfully pulled the enemy away from *Pathfinder*, which was the purpose of this dash to the edge of the system. Lieutenant Weaver sat at the tactical station listening to the weapons control officer who sat at his station in the Weapons Room, one deck down and directly under the bridge. His light auburn hair was matted with sweat as he valiantly coordinated the offensive and defensive missile launches. The wrinkles around his eyes, as they narrowed in concentration, showed that he was feeling the strain from the weight he carried. Brighton knew that if he continued to show no sign of worry, the crew would also feel there was no need to worry; at least, that's what he has been taught was "official command doctrine". He had often thought that premise undervalued people's imagination and intelligence. Jordan glanced over from his scan console every few minutes to ensure that the commodore was not worried, so perhaps the powers that be knew a thing or two. This was the reason Brighton let no trace of his inner turmoil reach his features. This did not stop the crew from having a high level of anxiety for this new, and very frightening, experience, but they would hide their fear, if for no other reason than the fact that their friends and comrades were not showing theirs.

This was the only service that Brighton could perform at this point in the engagement. It was also one of the most difficult tasks for him personally.

Unlike the crew, however, it was not fear that he felt burning through his core.

He studied his hands as the events of the last few months replayed themselves again in his mind. He pictured the cheery campfire and the ersatz terraforming team; the laughing and camaraderie that had been interrupted by gunfire and death. The imposters; pretending to be Warner citizens were in actuality, DaGaman Family Marines. Those troops that had taken the life of Drew Le Vesconte without thought or hesitation. The same troops who had taken the life of Jens Fujinami in a later battle and would have killed him and the rest of his crew to take *Vanguard* away from them. They had killed gentle, laughing, Fujinami, who always excitedly waited for another mystery to explore.

Those same murderous troops, or their accomplices, were following behind his ship, gaining slowly. They still thought that they could gain from the death of those that he had come to consider friends. The anger coursed through his system as he contemplated the situation. He looked again at the chrono that was mounted on the forward bulkhead. These red numerals had become his lifeline to reality. He watched as they slowly turned from one number to the next. He had a plan that would make them pay. If he was very lucky, their deaths would finally lay his ghosts to rest and he would no longer be troubled with the belief that he was the one who had led his friends to their deaths.

The bridge was shaken by another near miss. The targeting of these Forrest ships was not up to the accuracy standards of either the WSN or the Combined Fleet. This was not truly surprising considering the fact that there could not possibly be experienced officers on the bridges of those felonious ships. This was a function of the makeup of the Combined Fleet and the Families in general.

Following the conclusion of the Vector Rebellion, sometimes called the First Interstellar War even though there had yet to be a second war, the composition of the Combined Fleet was set by treaty. The four Families that had major holdings outside the Sol System were required to supply officers to the CF to provide the manpower to staff the ships that they were also required to supply for the common defense. As a safeguard to the peace, the crews for those ships were drawn from the citizenry of all seventeen of the Families, both space-faring and earthbound. Thus the officers and the crews would share a loyalty only to the Ruling Council and not to any single Family. The result of this treaty for the Forrest Family, however, was that they numbered no trained Fleet officers among their populace. They didn't need any because, according to the treaty, they had no warships to command. By law, they were allowed no fleet. They had no fleet. *Until now*, Brighton thought.

Brighton looked at the red numerals one more time. He did the calculations without conscious thought. It was amazing, he thought, how much easier it was to do certain tasks when you became accustomed to them. Before his *Vanguard* experience, he was content to run all calculations through the astrocomp. When he no longer had that crutch to lean on, he found that he not only could, but that he really *enjoyed* doing those calculations in his head. The ship lurched slightly as another missile was launched from their stern tubes.

Brighton was grateful that the officers of the ships behind them were as inexperienced as they were. He had feared implementing the plan to supersaturate the defenses of a single ship the way they had, even though that opened a hole in the closing net around them whereby they had extended this chase several more hours. A good officer would have recognized a good idea when he saw it, and would have done the same sort of thing to overcome *Yargus'* defenses and destroy her. Either the Forresters were not smart enough to recognize a viable tactic, or they were unfamiliar enough with their ships' capabilities to be able to figure out how to duplicate the feat. That was one reason Brighton had not ordered the same plan used on other ships, pulling them farther away from *Pathfinder* was enough to accomplish his mission.

Of course, a third possibility was that they were holding back in order to capture the ship instead of destroying it. Brighton would see to it that eventuality did not occur.

While the Forresters lacked the skill, or possibly the will, to do major damage to *Yargus*, their own return fire had proved ineffectual as well, though for a completely different reason. They had simply chosen not to use the firepower needed to saturate the defenses of the pursuing vessels. it was enough merely to slow down the pursuit. They were managing to do just that, but they also needed to conserve their munitions for the difficult part of the battle which was yet to come. Managing the ship's stores and munitions were the captain's responsibility. Managing the battle and the recovery of *Pathfinder* were his. He glanced again at the chrono. *Dagger* should be here by now if they had been able to get their jump engines working. It would appear that they were going to have to fight this alone.

"Captain," PO Dan Jordan called from his scan console, "Two of the destroyers are braking hard. It looks like they are headed back in-system."

Commodore Brighton leaned his head on his hand and quietly tapped his index finger on his chin as Ramirez turned to look him.

"We are running out of time. We can't wait any longer for *Dagger*," he said cryptically.

Several moments later, Brighton sat up straighter and, with a final glance at the chrono, said, "We will now begin. The Forrester ships are spreading along this front," he began, motioning with his hand at the screen to his right. We will turn directly into their formation and begin shedding velocity. I want you to fire all broadside weapons as they bear and continue the engagement with the forward tubes at the highest rate of fire you can manage. I want all fire concentrated on these two ships to the right of their formation. Make them believe that we want to go through there."

Ramirez nodded his head as he took in the plan that the commodore was laying out. Soon the irrepressible grin that had become his trademark was again evident on his face. "They will launch everything they have to keep us from getting through," he stated firmly but without the anxiety that would have accompanied the pronouncement a few hours ago.

"Yes, Captain," Brighton said as he leaned back and steepled his long fingers in front of him. "It would be a shame if they used up all of their munitions for no result."

Ramirez turned with a nod and moved to his command chair. He stood with his hand resting on the back and began issuing the orders that would put into motion the commodore's plan.

The captain finally turned to Brighton and said, "Course change and fire plan are ready, Commodore. Awaiting your command."

"Please proceed, Captain," the commodore said, with no trace of the rage that was consuming him. "Your jump coordinates are the same as our incoming point of entry." He then turned to his intercom and toggled it on. "Major Chowdhury, prepare for boarding action. Embark your Marines."

"Aye-aye, sir," came the immediate response. her tone was perfectly flat, which Brighton recognized as the only outward sign of anxiety he'd ever been able to detect in her.

At Ramirez' motion to the helm, *Yargus* started to turn. As the port broadside missiles left their tubes and the ship continued its turn, Captain Ramirez turned to Engineering. "Charge the capacitors on the jump engines," he said.

"Hits on target one," Jordan called from scan. "Tango-1 is falling out of formation. It looks like it has lost its engine. The other ships are shifting position to cover the gap."

"Multiple launches," Jordan called out excitedly. "Twenty, twenty-five, thirty, ...I make it thirty-four missiles, sir. Eighteen seconds to impact."

Ramirez turned casually and looked at Brighton. At the commodore's nod, he said, "Jump the ship."

CHAPTER 43
18 October
Warner Family HQ, Earth Orbit

Admiral Cosina sat at his desk , looking out the floor to ceiling windows at the gentle curve of the Earth as it spread below his view. Quito had moved past the terminator into darkness since the last time he had consciously noted its position. He tossed his stylus gently back on top of the stack of hardcopy reports he had begun reviewing several hours ago. He had only moved a single document to the outgoing bin on the corner of his desk. He sighed loudly. The admiral had come to that indefinable point where he realized he was not going to accomplish anything useful for the rest of the day. With the *Pathfinder* issues pushing out all other thoughts, he was finding it impossible to focus. There was not enough information yet to make further plans, but his mind continued to cycle through possible scenarios without data. He knew the problems with that sequencing and, again, decided not to color future facts with his current perceptions and guesses.

He made another attempt to focus on the top-most page in front of him, without success.

With a barely audible sigh, he spun his chair to look out his windows again. That didn't distract him from his current chain of thoughts, and he tried again to finish just this one document. The admiral had just restarted the report when his comm chimed.

"Sir, Lieutenant Griggs to see you," came his admin's voice when he had answered.

"Send him in," Cosina responded immediately.

Lieutenant Griggs entered the Admiral's office tucking his softcap into his belt, baring his head in his superior's office. He came to attention and snapped a salute, as regulations required, which the admiral returned.

"At ease," Cosina said as he motioned for the lieutenant to be seated, then reseated himself.

"When did *Avram* get in?" Cosina began, surprise evident in his voice.

"She docked fifteen minutes ago, sir. I came straight here," Griggs stated, sitting forward on the front edge of his seat. "Captain Carmichael shaved some time off our ETA. I didn't think the intel should wait, but I couldn't trust the security of our communications," Griggs replied, handing over a folio.

The admiral set the folio aside for his aide to transcribe later into his preferred hardcopy form.

Griggs began with a deep breath, "Commodore Brighton ordered me to report directly to you with a detailed summary of everything that had transpired up to the point where *Avram* separated from *Dagger.*"

"Very well, Lieutenant. Let's have your report," Cosina said, placing his hands on the desk. "And don't leave out any details, no matter how insignificant,"

Griggs began with the trip to Betre, and the engine trials on *Yargus*, then detailed everything they had uncovered and everything they had done since leaving Earth two months before. He didn't add any personal observations, until Cosina asked him for them after he had finished.

"And the prisoners?" Cosina questioned, obviously still making mental notes.

"They are being transported to a secure facility as we speak, sir," Griggs answered.

He wasn't used to speaking directly with any admiral, let alone someone with the legendary reputation of Admiral Cosina. Griggs wasn't entirely comfortable, and he kept telling himself that the quicker he got Admiral Cosina all of the information, the quicker he'd be back into his comfort zone.

Cosina went silent and steepled his fingers with his elbows on his chair's armrests. He leaned back slightly in his chair, absorbing what had been said and connecting it to what he already knew or surmised.

Lieutenant Griggs waited quietly, expecting further follow-up questions that never materialized.

Abruptly, the admiral stood, and Lieutenant Griggs snapped quickly to his feet as well, in automatic response.

"Come with me, Lieutenant," Cosina said, striding out of his office.

Griggs followed, a little dumbfounded. Admiral Cosina moved briskly and Lieutenant Griggs was hard-pressed to stay on his heels, and replaced his cap as they moved quickly down the corridor.

The admiral never looked back, secure in the assumption that the lieutenant was closely following, per his orders. Shortly after rounding another corner, they reached the main tube. Lieutenant Griggs was unclear where Admiral Cosina was headed, but he followed him into the lift. As soon as the tube came to a stop and the doors opened, the admiral was moving again. Admiral Cosina clearly knew the way to wherever it was he was headed. Lieutenant Griggs finally realized they were in the central hub, and had his first subconscious warning that he wouldn't be any more comfortable when he arrived where the admiral was headed, when they walked up to an administrative desk only to be motioned past as the admin staffer took in the look on the admiral's face.

Large, ornate, double mahogany doors opened before the burly admiral reached them and Cosina strode purposefully on into the room. Lieutenant Griggs followed tentatively, as he had never been in this section of the central hub before.

"Gerry," Cosina began, as his friend looked up from behind his massive desk, "Lieutenant Griggs just arrived on *Avram* and brought me Brighton's report. He brought us additional evidence from a DaGama base they found in the Antoc system and documents they were able to recover. Those documents tie both the DaGama Family as well as the Forrest Family to the theft of *Pathfinder* and the attempts to secure *Vanguard* as well. Lieutenant Griggs, please tell our CEO what you told me,"

"Umm…, sir? Yes sir. Start at the beginning sir?" Lieutenant Griggs said, coming to attention and pulling off his hat as he belatedly realized where he was.

"That's generally a good starting point, Lieutenant," Admiral Cosina said, allowing a little mirth to seep out at the young officer's obvious discomfort. "In this case, however, just the information from the asteroid base."

"Yes, sir," Lieutenant Griggs said as he was able to find his verbal stride again, and walked back through all the pertinent details. For his part, Warner patiently listened, only asking occasional clarifying questions.

Once Griggs had completed his report, Cosina immediately rounded on his CEO with determination.

"We have corroborating proof on much of the interference we've known about. The prisoners are being secured and we have all of the data that Brighton's teams have gathered. So far, we haven't found any proof that Walton has directly interfered with our shipping, but we still have no reason for their unusual construction deviations. It's time for action, Gerry," Cosina stated matter-of-factly.

"We have *some* proof, Conrad, but I am not sure we have *enough*. Additionally, we have no way to know what the situation is right at this moment. We can't escalate this further without notifying the Ruling Council,"

"So, what would you have me do in the meantime?"

"We have already sent a sizable force to Tannar. Dispatch Admiral Koutsoudas with *Victory* to join them there. We don't know what the situation is in Worth, but that seems to be where things are happening. We know that Brighton pursued *Pathfinder* there, but we have no proof that ship is still there. Notify all systems to move to a higher state of alert and readiness.

Also, let's make an effort to get all ships out of their current maintenance cycles and reinforce Gateway and our Earth offices. I don't want anyone to get ideas with *Victory* out of the system."

"Meanwhile, we gather every shred of evidence that we have available and the three of us are going to notify the RC as soon as we have enough evidence to back up our claims," CEO Warner said while moving to his desk and reaching for his comm.

Lieutenant Griggs felt his heart skip as he heard the words "the three of us" in association with the notification to the Ruling Council. He was thankful that his black Warner Fleet uniform couldn't show the amount of perspiration it was absorbing.

CHAPTER 44
18 October
Forrest Main Complex – Earth

Amanda Forrest stabbed the last bite of veal parmesan and raised it to her lips, but did not really taste it as she chewed. She continued staring at the report in front of her on the screen, then realized she had gotten distracted by a stray thought, lost her place, and had to backtrack to find it. She kept scrolling back, but couldn't find anything she remembered having read. When she arrived at the top of the first page, she didn't recognize that either. Just how long had her mind been drifting?

James came in to clear away the dishes, knowing the instant she was finished. She leaned back to let him collect the soiled china, wiped her mouth with the linen napkin and added it to the top of the stack before he got out of reach. She wished, absently, that all the problems the Forrest Family was facing could be cleaned up and whisked away as easily.

The door had barely closed from James' exit before it opened again to readmit him. "Ma'am, Mr. Blasingame is here to see you. He says it is a matter of some urgency."

Based on the timing, Amanda deduced that James had not deemed it sufficiently urgent to interrupt her meal. "Let him into the study, James. I'll join him presently."

"Very good, ma'am."

The CEO was true to her word, taking only long enough to check her appearance and smooth her features before walking into her study. Blasingame rose from a comfortable leather chair as she entered and took her hand warmly when it was offered. That was all the time either of them felt like wasting on pleasantries.

"What is it?" Amanda asked before she had even seated herself at the desk.

"Two things, one internal and one external, and I'm not sure which is the more important." That admission startled Amanda more than Blasingame coming to her home to deliver news.

Stavros Blasingame had been what she considered her "spymaster," regardless of his actual job title, for fifteen years. In all that time, he had delivered information face-to-face perhaps a dozen times. Being unable to prioritize the information for her consumption had happened exactly never.

"Give me the internal information first, then," Amanda directed. She thought it might be a nice change from worrying about how their defensive buildup was lagging, a topic that had never been far from her thoughts for weeks now.

"Matthias is making a play for your job; not just plotting, but he's made his first move." Amanda's eyebrows rose at that. She had known of Matthias' ambitions, of course, since her brain cells were still receiving oxygen, but she'd thought they'd had an understanding that *now*, of all times, everyone needed to be pulling together in order for the Family to survive. The fool!

Blasingame continued, "Orders went out of the Off-Planet Affairs office early last week, above his signature, removing all authority from Epi Solomon and turning everything over to the military. Following the obvious thread, I was able to locate documented evidence of an agreement between Mat, Fleet Admiral Choi, and General Hart to support him in exchange for a seat each on the board."

"When did the collusion start?"

"Mat approached them eight or nine months ago, but he was not able to convince them until recently, again, by citing the deteriorating situation on Granada and appealing to their sense that if the military were in charge, the problems would be resolved at once."

Either Blasingame was oversimplifying their thinking, or they were even bigger fools than Mat. Knowing Blasingame, the men were most likely the fools he was implying.

"Well, the timing and the actions are a bit surprising, but the fact that he would eventually make his move is not. I have a number of counter-moves ready to launch. It will just be a matter of deciding how big a hammer to hit him with. What news on the external stage, then?"

"Gerry Warner has arranged for himself and an unknown number of guests to be added to Rial's schedule tomorrow. He's also called for an emergency meeting of the Ruling Council the day after."

Blasingame's voice was calm and measured, as it always was, but the words cut through Amanda Forrest like a scythe. Her mind stopped working for several seconds, and for many more the only thought she could form was, *No. No, it's too soon. We're not ready yet.*

She tried hard to avoid giving away her feelings of dread, but she knew Blasingame would spot them anyway. "How much does he know?" she asked finally.

Unfortunately, this was not an eventuality for which she had countermoves ready to launch. The countermoves had been planned out, of course, years before, but every detail had to be managed in its proper order, and they wouldn't be ready to move directly against Warner for almost another year. The way the schedule now stood, fourteen months wouldn't quite be enough.

She needed time to figure out what she could do to hold things together a little longer. What actions could she take to stall while they mobilized? She needed time!

Blasingame understood the true question she was asking and shook his head slightly. Rather than answer her immediately, he rose and moved to the far end of the room, poured two drinks and brought them back. Amanda was grateful for the respite, and she used the time to organize her scrambled thoughts. Only after both of them had sampled the liquor did he speak. "I would have thought Warner knew very little. In fact, I would have thought they knew nothing more than that *Pathfinder* had been taken. That's part of why it's difficult to know how much importance to give Warner's actions. I don't know what they know, and not because I haven't tried to find out."

"I'm not accusing you—"

"I know. Still, it's disquieting not to have more information on which to base a response. I can positively say two things, though: First, it should be impossible for them to know of our involvement unless they found out from DaGama, which they would not do without implicating themselves, and even then would be a case of their word against ours. And second, they have to know more than we thought they did, or their current actions would be a disproportionate response." He sat back and took another sip from his glass.

"Gerald Warner is not one to jump the gun," Amanda replied, not because Blasingame didn't already know that, but because it was something to say while her mind tried to piece together a course of action. He didn't bother to nod in agreement. "What time is he meeting Felix tomorrow?" she asked.

"Ten a. m. Felix' time."

"And what time Tuesday for the Council meeting?"

"It's not scheduled yet, but I would predict one or two in the afternoon."

"Is there anything else I need to know?"

"No. Is there anything else you need me to find out? Apart from what Warner does and doesn't know, of course."

The attempt at humor wasn't enough to lift her spirits. Even if the answer were that Warner knew nothing at all, there would still be too much scrutiny for some plans to continue moving along, in the short term. "Nothing with the same priority, Stavros. I need to know what they know, what they can prove, even what they suspect, if I'm going to have my defense ready. Fortunately, I may have a way to get some of that knowledge. I'll let you know if I turn up anything."

Blasingame recognized his cue to exit, so he tossed back the rest of his drink and rose. He turned back at the door and nodded a good-bye, but Forrest already had her comm set plugged in and only nodded in return.

"Noelle," she said into her mouthpiece, "get me Felix Rial." Noelle went to work at once, but there was still a delay before she could get Rial in the link. The Rial sigil filled the comm screen, a steel-blue eagle on an orange field. Amanda used the time to make a list of projects that would have to be accelerated, those that would have to be put on hold, and there were two that would have to be dropped altogether. One of these last, she noted with a smile, was a military operation where word of cancelation would have to come from OPA. Wasn't that a shame that General Hart's reward for supporting Mat would be to have Mat cut her out of the biggest planetary invasion ever. Yes, simply too, too bad.

"Rial." The flat voice interrupted her thoughts. His scowl made it clear he wasn't any happier about being pushed into a corner by her than he was the last time they had conversed.

She didn't try to put any false lightness into her own voice. She hated the duplicity such a move implied, and besides, it wasn't likely to do anything about how either of them felt about the other.

"Felix. I understand you've been talking with Warner."

"Yes," he said in the same tight voice. Rial could be as diplomatic as a head of state needed to be, when he wanted. Like Amanda, though, he didn't waste the effort when there was no need to keep the person you were talking to from knowing just how low you thought they were.

"I need to know what he wanted," she said. There was no response.

"Felix, what did Gerry talk to you about?" Amanda asked.

"He was invoking the seven-two privilege to convene a general Ruling Council session." He didn't add anything else. It was like talking to a machine; obedient, but not helpful.

"Did he say what it was about?"

"No."

She waited, but he didn't elaborate. Oh, the man was infuriating! Still, if Warner had exercised the powers outlined in section two of article seven, then there was only one thing it could be about: Warner was going to level formal charges against another Family.

"Did he say which Family would be accused of wrongdoing?"

Rial looked surprised at the question. "No, he didn't," Felix finally said. "Based on our earlier conference, I guess I had assumed it would be Forrest, but Gerry never actually said. I imagine he will fill me in on details tomorrow."

"I'd like a copy of those details, Felix."

The stiff demeanor returned at once. "I imagine you would, Amanda." He held her gaze for several moments before looking away. His posture relaxed slightly. "All right. It's not as if I have any choice."

Forrest's smile was cold. "No, it's not. Good night, Felix. We'll talk again tomorrow." Rial signed off without another word.

Amanda leaned back in her chair and let out a deep breath. She wasn't really as cold as she pretended to be. It was a useful mask to put on, however, and something that she frequently used in her line of work.

She rose and walked back to the credenza, refilling her glass and taking it back to the desk. The contents disappeared in two swallows, then she reached out and dialed a directory code from memory. When Matthias' face filled the screen, Amanda's mask was in place; the one which showed tension but not suspicion, the "you-have-nothing-to-worry-about-from-me" mask.

"Mat, so sorry to comm this late, but there are some serious changes we need to make, and quickly."

"What is it, Amanda?"

"I don't have enough details yet, but the balloon may be going up much sooner than we planned on. I need to meet with you and your team in the OPA operations center at six a. m. We're going to have to rework most of our contingency planning, and quickly."

Mat looked confused. "What are you talking about? What's happened?"

"Are you ready for a war? Today? Right now?" Amanda regretted saying it, because after the words were out, she knew he was thinking that she had meant him, personally, and not the whole Forrest Family together. She should have been more careful not to put him on the defensive, since she still needed him thinking straight. "Warner has called a seven-two for the day after tomorrow. I don't know how bad it is, but if they think it's serious enough to take it to the Council, we need to be better prepared to respond to whatever accusations they make."

Mat's eyes grew unfocussed while he thought, and she could tell that her last statements made him more relaxed, rather than the opposite. *Blasingame was right*, she thought, though she hadn't really doubted.

"I'll have everyone there, Amanda. If there's nothing else, I need to make some calls," he said earnestly.

"If I think of anything else, I'll bring it with me in the morning. Good night."

"Good night."

After she switched off, she made a mental list of who she might nominate to fill Mat's seat next month, or maybe December.

Sometimes, the coldness was not in the mask.

CHAPTER 45

18 October

Yargus / Pathfinder

"Okay, listen up," Major Sheli Chowdhury yelled to her troops on Shuttle One as they moved down the central walkway to settle into their positions. "As soon as we come out of jump, we are going to drop away from *Yargus* and head for pod three on the port side of the target. We will attach quickly and go in hot. Anyone not in 'Warner Black' is to be considered hostile whether they are in military uniform or civilian dress." Several heads came up at this change from the normal rules of engagement. "Any Warner that you encounter may be friendly, or they may be assisting the enemy, so take all precautions to ensure your safety. This is a possible hostage situation, but don't make any assumptions."

She looked down the double row of Marines. There were four groups of six each in two facing columns. Most of the troops were sitting casually, looking up at her. A few were more rigid, some lounging, as much as it was possible to do, while in full armor. Most had their helmets racked under their feet, but a few had them latched on their heads and a couple even had the face shields down. This was not a common practice, but each person prepared for combat in their own way. "I know that most of you don't know me, except possibly by reputation," Chowdhury continued, "but I want to make something perfectly clear." She waited until every eye, or face shield, was turned her way. "No one takes unnecessary risks. I run a tight team, and I know I can be a real pain sometimes, but you have not experienced pain until you try to hotdog in my outfit. There will be nothing left of you but a puddle that dribbles out when I pop your armor release. You are going to envy Humpty Dumpty. Hear me on this. Understood?"

"*Ooh*-rah!" they shouted back in unison.

"All right, Lieutenant Mdembe is going to be coming in opposite us with his team on the starboard side of the target, so watch your fields of fire. My team will head forward and take the bridge and Lieutenant Kelley's team will take and secure Weapons Control. Lieutenant Mdembe and his team will go aft and secure Engineering and also the attached shuttle. Any questions?"

As she spoke, she felt the slight nausea that indicated *Yargus* had jumped. She knew the plan and her two shuttle pilots had been informed, so she wasn't surprised to feel the vibration in her feet as the shuttle engines went hot. They were still in standby so there was no appreciable noise to deal with, but they were ready as soon as Commodore Brighton gave them the 'go'.

"Saddle up, troops, we're going hot in about thirty seconds."

She made her way several steps forward to the side-facing jump seat that would be her perch for the flight in. To her right, she had a view of the flight deck and the two shuttle pilots. They looked young and nervous and Chowdhury knew this was their first combat insertion, but they were competent and they hid their nerves in the routine of their flight checklists. To her left, she could see her Marines doing their own final checks on their gear and many more helmets were being secured in place. It was amazing how soothing the routine was. It calmed her at a time when she needed it most.

The memory of her exit from *Pathfinder* still stung. She couldn't believe she had let those idiots get the drop on her in the corridor; but even worse, she couldn't believe she hadn't seen it coming. Brighton obviously had, and was able to make some preparations, but she had been so concentrated on finding the mole that she had begun to relax after Jhonsruud had been captured. It was inexcusable. She knew her nerves were ramping up in anticipation of evening the score, but she had to fight down that feeling and concentrate on doing her job. Everyone needed to concentrate.

The pilot's strong soprano cut into her thoughts before she had time for any more self-recriminations. "Drop in five, four, three, … Drop." Chowdhury felt the release of the shuttle from its attachment points on *Yargus*. She could see the stars stream across the forward view plate in front of the pilot as the shuttle raced ahead to its rendezvous with her old ship. She watched the scan console that showed *Yargus* continue to accelerate toward Granada where they would use the planet to slingshot themselves at a faster velocity out away from the gate they had just used to reenter the system. She could imagine the consternation and surprise on the faces of the crew who probably thought they were jumping to safety, only to find themselves back in the same system they just jumped from. *Yargus* was now inside the fleet that had been chasing them and that fact had given them a chance to get onto *Pathfinder* without the Forresters intervening.

Conscious of her previous thoughts, she dismissed *Yargus* and concentrated on her part of the mission.

The pilot was very good and was pushing through *Pathfinder*'s weapons envelope at the shuttle's highest rate of acceleration. At the last possible second, she flipped the shuttle and began the deceleration that would put them next to *Pathfinder* in a position to attach and board. The pilot knew the most dangerous part of the mission was this one, when the enemy could detect them and engage them before they got close enough to attach. She was using her high speed to minimize the time *Pathfinder* had available in which to engage them. The element of surprise appeared to be working in their favor, since the target's weapons remained inactive.

Major Chowdhury watched as her old ship's image grew until it nearly filled the screen in front of the pilot. The view had not changed, regardless of the orientation of the shuttle to the target. Still there seemed to be no reaction from the old destroyer. The pilot was excellent and the shuttle came to rest mere meters from the target hatch, well inside the envelope where none of *Pathfinder*'s weapons could bear until the ship moved.

"Breach team to the airlock," Chowdhury called without moving. The four Marines in the front seats jumped up immediately and, grabbing their gear from the compartments around them, moved to the airlock between Chowdhury and the rest of the troops. A solid hatch came down just in front of Chowdhury, cutting her off from the flight deck. She looked back to see that all of her Marines had donned their helmets and closed the face shields. She closed her own and said, "Seal check." Twenty armored thumbs went in the air and she keyed her radio to the flight engineer. "Vent the compartment." She knew all the operational timing was in her hands now. Lieutenant Mdembe would not breach the inner door on hold ten, which was on the opposite side of her target airlock, until forty-five seconds after she entered the boatbay. She wanted to give the defenders plenty of time to focus on her before the other team came in at their backs. She would rather have had Bravo Team come in through the hull behind the bridge and work aft, but they needed to minimize the damage to the ship if they were going to be able to grab it and jump out before the Forrest fleet could return from the outer system. *Yargus* would fight to keep the Forresters off their backs while they made whatever minor repairs were necessary, but they couldn't delay them indefinitely.

The breach team hit the override on the airlock and opened the inner door while she listened to the high whine of the massive pumps that were pulling all the air from the cargo area where the troops sat.

When the inner indicator light turned amber, her breach team hit the release on the outer lock and let the slight decompression carry them out into the void. Knowing they knew their jobs, Sheli left them to it. She waited to a count of ninety, then she stood and moved to the outer hatch. A collapsible, temporary tube was now connected between her hatch and the outer hatch of the airlock on *Pathfinder*. Her team fell in behind her at a silent signal and Lieutenant Kelley with his team behind that. At her silent count of forty-three, she launched herself lightly out the hatch and moved through the tube, grabbing the occasional handhold to keep herself oriented properly. Just before she reached the end of the tube, she checked that the hatch was clear before she sailed, feet first, through the open hatch and into the cargo hold. As soon as she passed the threshold of *Pathfinder*, its gravity began to assert itself and she touched down easily in a maneuver that appeared graceful, but was truly one of the most difficult skills that she had ever been able to master. She moved forward to the inner door as the breach team sealed the tube behind the last member of Kelley's team. The air was quickly being replaced in the cargo pod from cylinders brought along for that purpose. She reached the inner hatch as the breach team attached their charges and, at their look, she signaled with a nod and the hatch blew inward. She checked behind her at the rest of her team, then primed her blast rifle and leapt through the hatch into the boatbay. She rolled, as well as you could roll in Marine PUMA armor, and came up tracking from aft to fore and saw no immediate resistance. She heard two more of her team land behind her and mirror her actions, each with their own zone of responsibility. She stopped dead and yelled, "Hold fire," as she saw and identified the four figures in rumpled Warner Fleet uniforms standing at attention by the forward bulkhead, saluting her.

"Welcome home, Major," said the young lieutenant in front. "We've been expecting you. We could really use your help."

ABOUT THE AUTHORS

Jeffery L. Cheney
Jeff is the second of the seven Cheney brothers. He has worked as a civilian contract mechanic for the US Army, a heavy equipment mechanic, a high school teacher, and currently works in high technology computer chip manufacturing. Jeff has been writing science fiction and fantasy stories for enjoyment for over thirty-five years and has published two SF novels with his brothers; <u>Dead Reckoning</u> and <u>Day of Reckoning</u>. <u>Force of Reckoning</u> is his third novel with his brothers. He is also completing his first solo novel, <u>Forged by Betrayal</u>.
He enjoys coaching youth basketball, working on cars and doing woodworking when the time allows.
He has three grown children and he lives in a small town in NW Oregon with his wife of 28 years.

Craig J. Cheney
Craig is the fourth of the Cheney sons. He holds degrees in Accounting, Business Administration, Computer Engineering, and Electrical Engineering. He has worked as a disk jockey, put on trade shows, organized a circus, taught classes on Shakespeare, Math, Debate, and Parliamentary Procedure, and is currently dabbling in rocket science.
Craig was the runner-up for the 2009 Next Mark Twain Award. He, his wife, and their children live on Utah's Wasatch Front.

Jared L. Cheney
Jared is the youngest of the brothers. He has worked for many years as an IT Executive, and is currently a Senior VP for a global cloud services company. He loves to travel and has lived and worked all over the US and in over 10 different countries. He holds degrees in Business Administration, Information Systems, and Electrical Engineering.
Jared and his wife live in the Portland, Oregon area with their children.

The authors all graduated at or near the top of their respective classes at the same high school on the Oregon Coast. All three are Eagle Scouts and volunteer their time to support The Boy Scouts of America.